Prairie Dogs

Prairie Dogs

COMMUNICATION AND COMMUNITY IN AN ANIMAL SOCIETY

C. N. Slobodchikoff

Bianca S. Perla

Jennifer L. Verdolin

HARVARD UNIVERSITY PRESS

Cambridge, Massachusetts, and London, England · 2009

Library of Congress Cataloging-in-Publication Data

Slobodchikoff, C. N.
 Prairie dogs : communication and community in an animal society /
C. N. Slobodchikoff, Bianca S. Perla, Jennifer L. Verdolin.
      p. cm.
 Includes bibliographical references.
 ISBN 978-0-674-03181-4 (cloth : alk. paper)
1. Prairie dogs. 2. Animal communication. 3. Social behavior in animals. I. Perla, Bianca S.
II. Verdolin, Jennifer L. III. Title.
 QL737.R68S56 2009
 599.36'7156—dc22   2008016573

Contents

Preface

When each of us started our study of prairie dogs we were focused on their social behavior and communication. The complex issue of prairie dog conservation was not at the forefront of our study emphasis or thinking. However, as we worked at our field sites, learned more about our prairie dog "subjects," and talked with others studying prairie dogs, the plight of these animals and the tremendous need to conserve them became extremely clear.

Drawing on our diverse backgrounds in ecology, conservation biology, and animal behavior, and combining them with our common experiences studying prairie dogs, we explored the issue of prairie dog conservation together in many informal discussions. Ironically, we realized that a major part of the problem contributing to prairie dog decline dealt with communication—precisely what had brought all of us together to study prairie dogs in the first place! Of course this lack of communication was not between the prairie dogs themselves, but rather between humans, and between prairie dogs and humans.

The first step in conserving anything is an awareness of the value of its existence. Without getting to know an animal, plant, or person, one can intellectually respect its right to life but one cannot truly feel its worth or know what is required to save it. Without this appreciation, the effort required for conservation is particularly difficult to muster. When the idea for this book was formed, there were no comprehensive books that brought together all aspects of information on prairie dogs and their interactions with humans. We started writing this book as a form of communication, a window into the life of prairie dogs that allows people to know them better and, we hope, will ultimately encourage more informed participation in the effort to stop prairie dog decline.

As researchers of prairie dog behavior and communication we look at prairie dogs differently from wildlife biologists and ecologists who are ordinarily charged with the responsibility of managing and conserving species. For this reason, we bring a different perspective and new information to the issue of prairie dog conservation. For example, we see value in prairie dogs as individuals as well as populations, and we value prairie dogs as animals that can teach us much about sociality, evolution, and communication in non-human societies. Because the questions researched in animal behavior and communication at some point always lead to questions about ethics, intellect, and self-awareness in animals, we deal more with ethical considerations of prairie dog conservation than a standard wildlife management or conservation biology text. We also present more detailed information on the uniqueness of prairie dog communication and behavior and show how it is relevant to conservation issues.

Because conservation of a species is such a multifaceted endeavor, we explore the ecology, economics, and population biology of prairie dogs. Our goal is to be synergists bringing together information on prairie dogs to communicate more fully their essence and value to humans and other species. To this end, we present chapters with extensive references on all aspects of prairie dogs and subjects relevant to species conservation.

In this book we are not dealing with the question of whether or not prairie dogs should be conserved: we are already convinced that they should be. This book is an attempt to gain a clearer and more comprehensive understanding of who prairie dogs are by incorporating new information about their communication and sociality alongside standard information about populations, economics, and biology. We are interested in exploring what roles prairie dogs play in enriching grassland ecosystems and human societies, what risks are present to human societies and other grassland species if prairie dogs go extinct, and what avenues and obstacles to their conservation exist.

We have made an effort to break the issue of prairie dog conservation out of its traditional and fairly isolated frame in search of more creative solutions to prairie dog decline. We do this initially by looking at prairie dog conservation from the angle of animal behavior and communication. We also connect prairie dog issues to larger patterns, obstacles, and solutions occurring in the wider sphere of biodiversity conservation. We relate prairie dog decline to common patterns of biodiversity decline and explore cutting edge solutions from other environmental science disciplines that have not been traditionally applied to prairie dog conservation. Thus, our intended audience includes not only those who are experts in prairie dog biology, ecology, conservation, and management, or naturalists and lay people with a curiosity about prairie dogs; but also researchers in animal behavior who are interested in how their knowledge can inform conservation, and students of conservation biology who are interested in learning about patterns, conflicts, and solutions in species conservation that are distilled into the tale of one genus.

We would like to thank the following people in helping us with the creation of this book. We could not have done it without their support. We thank Ann Downer-Hazell and Vanessa Hayes of Harvard University Press, and Richard Reading for extensive comments on the manuscript. We also thank James Detling and James Hare for helpful comments on the entire manuscript. A number of people provided helpful information or reviewed sections of the manuscript. For this, we thank Dean Biggins, Ana Davidson, Judith Kiriazis, Lindsey Sterling Krank, Lauren McCain, Stephanie Nichols-Young, Bill Van Pelt, Jonathan Proctor, Erin Robertson, Nicole Rosmarino, and Steven Travis. We also thank Allison Kennedy Taylor for editing sections of the manuscript. Our gratitude goes to our family members for their support during this endeavor, and to the prairie dogs.

Prairie Dogs

1

Prairie Dogs and the Big Picture

This is the story of prairie dogs (*Cynomys* spp.). Driving through the American West, you get a glimpse of them, sometimes right by the side of the road. Larger than other ground squirrels, smaller than woodchucks or marmots, these unique creatures are part of the western grassland landscape and part of the fabric of its history. But, in a larger sense, the story of the prairie dog is also the story of every other animal species living on earth.

Traditionally, prairie dogs have been targeted as pests and as vermin. Attempts at deliberate and systematic eradication have taken their toll. Now, instead of numbering in the billions as they once did when buffalo walked the prairies, prairie dogs are on the brink of extinction. Yet they have much to teach us about the complexity of animal life and the interdependence of ecological relationships. Prairie dogs have a significant role in grassland ecosystems. They have a rich social life with many mysterious interactions that go on underground and out of our sight. And they have a sophisticated communication system that might outstrip monkeys and apes in its complexity, a system on the verge of language.

Why does it matter if prairie dogs go extinct? After all, lots of species have become extinct—the fossil record is filled with species that no longer roam the earth. We no longer have dinosaurs munching on plants at the edges of lakes, or pterodactyls flying through the skies. Among the mammals, saber-toothed cats no longer stalk their prey, Irish elk no longer lift up huge racks of antlers, and pigmy elephants no longer walk through dark forests. Aren't we just seeing the evolutionary process in action? Some species adapt to changing circumstances and survive, while others fail to adapt and die.

In geologic time, species do go extinct. However, such extinctions appear to have taken place over millions of years, often because of long-term changes in climate or catastrophic events such as collisions with large meteors or comets. We humans need to realize that we don't know how the extinction of the prairie dogs, happening in what is considered a blink of a second in geological time, will affect the grassland ecosystems and the health of our planet—and our survival as a species depends on the ecological health of our planet.

In this book, we explore the life of these animals. However, this book is not just another detailed and isolated account of a particular species. We are learning that to understand a species fully we must also consider its place in the larger ecosystem and its historical and current interactions with human society. Today, more than ever, we are increasingly aware that the health and fate of both human and non-human species are tightly intertwined, and that a conversation about one species often leads to a conversation about many others.

Conflict of Interest between People and the Natural World

Like so many other stories, the story of prairie dogs now hinges upon a conflict of interest between humans and the natural world. The North American prairie was once a vast, open land stretching from southern Canada to northern Mexico and from the plains states to the eastern edges of Arizona and Utah. Historically and ecologically, there is nothing else in the world like the great North American prairie, unbroken, immense, lying beneath miles and miles of open sky. Today, little of this natural heritage remains. Not much more than 150 years ago, grasslands were the dominant vegetation across North America (Henwood 1998). Now less than 1% of tall-grass prairie and 20%–30% of short-grass prairie remain, a decline caused by urban development, the expansion of agriculture, the development of water sources, and exploration for minerals (Gauthier et al. 2003). Grasslands have diminished so much they are considered to be an endangered ecosystem. In our drive to expand our economy, we have practically driven the grassland ecosystem over an ecological cliff.

The loss of prairie dog habitat is equivalent to the habitat loss that grizzly bears and wolves have experienced in the past century (Van Putten and Miller 1999). Before the year 1900, prairie dogs inhabited some 2%–15% of the Great Plains, with colonies often occupying more than 20,000 hectares (50,000 acres), and some colonies inhabiting 2 million hectares (5 million acres) (Knowles et al. 2002; Proctor et al. 2006). Prairie dogs now occupy

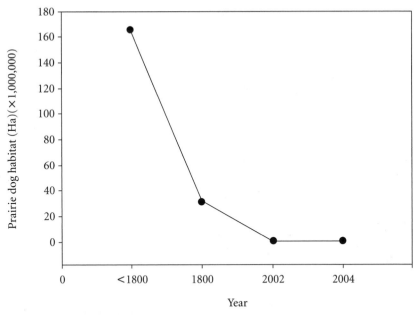

Figure 1.1. Decline in prairie dog habitat from prior to 1800 to 2004. Data are plotted from Proctor et al. 2006.

about 1%–2% of their former range, inhabiting only 600,000 hectares (1.5 million acres) of their original 40–100-million-hectare (100–250-million-acre) habitat (Marsh 1984; Anderson et al. 1986; Proctor et al. 2006) (Figure 1.1). But it is not only the prairie dogs that are declining. A plethora of grassland species throughout all trophic levels are waning as well, and this mass decline is a large threat to the integrity of the grassland ecosystem as a whole. Currently there are 55 grassland species that are listed as threatened or endangered under the Federal Endangered Species Act, and over 728 other species are candidates for such listing (Flores 1996; FR 67:40657–40679).

Today there are very few large prairie dog complexes left. In fact, when the United States Fish and Wildlife Service wanted to reintroduce the black-footed ferret into the wild, it was hard pressed to find prairie dog complexes big enough to sustain populations of predatory ferrets (Miller et al. 1996). This causes a problem for the black-footed ferret reintroduction program because recent reports indicate that the success of such a program hinges upon finding a greater number of large prairie dog complexes (Proctor et al. 2006). The largest remaining black-tailed prairie dog complex at Janos in northern Mexico decreased in size from 41,000 hectares (90,000 acres) in 1996 to about 23,000 hectares (50,000 acres) in 2000, primarily because the land was converted to agricultural uses.

One source of conflict between prairie dogs and humans has been over the issue of grass. Because prairie dogs eat grass, people in the past assumed that prairie dogs competed with cattle for forage, and were pests. C. Hart Merriam, a biologist who developed the "life zone" concept in ecological theory and was the first director of the Biological Survey, pronounced in 1902 that 32 prairie dogs eat as much grass as a sheep, and 256 prairie dogs eat as much as a single cow (Merriam 1902). Exactly how he came up with these numbers has never been clear. However, the view that prairie dogs compete with cattle has persisted for a long time, even into the present. Experimental evidence suggests that this view is either wrong or is not entirely correct (Detling 2006; Derner et al. 2006), but traditions die hard. One study of cattle foraging in some pastures without prairie dogs and others with black-tailed prairie dogs found that there were no statistically significant differences in weight gain between the two groups of cattle, suggesting that prairie dogs and cattle don't compete for the same food (O'Meilia et al. 1982). In fact, black-tailed prairie dogs can alter the plant composition of a site by allowing some plant species that livestock eat to become more abundant, actually favoring cattle (Bonham and Lerwick 1976).

There is even some evidence to suggest that prairie dogs do better in habitats that have been severely overgrazed by cattle. When cattle are not allowed to graze in grasslands, cool-season grasses such as wheatgrass and needlegrass increase in density, yet prairie dog numbers decline (Snell and Hlavacheck 1980; Uresk et al. 1982; Sharps and Uresk 1990). This decrease in prairie dog population relative to the decrease in grazing cattle has been proposed as a method to control the densities of prairie dogs, but this theory has been mostly ignored (Snell and Hlavacheck 1980; Cable and Timm 1987). In fact, prairie dogs tend to move into areas that are overgrazed by livestock, rather than stripping the vegetation before the cattle could eat it. When livestock heavily use or even overgraze grasslands, prairie dogs tend to increase in number. They occupy disturbed sites around cattle tanks, ranches, and other places that have been grazed (Knowles 1986), areas where there is a considerable amount of bare soil.

Perhaps the bare soil provides the prairie dogs with access to seeds buried in the ground. These seeds are packets of high-energy food that might allow the prairie dogs to reproduce and enable them to have a higher number of pups that survive (see chapter 3).

Historically, black-tailed prairie dogs coexisted with millions of American bison (buffalo) without any apparent competition. Bison numbers have been estimated at 30–60 million before the 1800s, and prairie dog numbers have been estimated at 5 billion (Fahnestock and Detling 2002). Like cattle, bison preferentially feed on grasses (Van Vuren and Bray 1983). They prefer to forage on the edges of prairie dog colonies (Kreuger 1986; Detling 1998), primarily because the clipping of vegetation by the prairie dogs produces forage that has greater concentrations of protein and higher digestibility (Coppock et al. 1983a, 1983b). Prairie dogs also coexist with pronghorn antelope, who prefer to feed on the forbs that are more characteristic of the centers of prairie dog towns (Kreuger 1986). Prairie dogs and large herbivores seem to coexist quite nicely when the grassland ecosystem is not disturbed by overgrazing.

Another source of conflict is the use of space. Prairie dogs tend to live in places that are flat—perfect places for subdivisions and shopping centers. As demographic changes occur in the United States and people move into the less populated areas of the American West, they come into areas that are inhabited by prairie dogs. Rather than finding ways to coexist with these animals, people exterminate them because it is often the cheapest and easiest approach to the situation. Many people feel that prairie dogs are merely rodents, so what possible difference can it make?

But it does make a difference. From an ecological standpoint, prairie dogs are considered a keystone species (see chapter 6), a species that widely affects other species in the grassland ecosystem. By digging burrow systems, prairie dogs affect soil nutrient cycling, plant and animal diversity, and small mammal abundance (O'Meilia et al. 1982; Agnew et al. 1986; Whicker and Detling 1988). Prairie dogs help support an entire ecosystem of predators, prey, insects, and plant communities. More than 160 vertebrate species are either casually or intimately associated with prairie dog towns, relying on prairie dogs for habitat and food resources (Clark et al. 1982; Kotliar et al. 1999). The widespread reduction of prairie dogs has had a domino effect on a host of species, causing some to become threatened or endangered. Several species have demonstrated a corresponding decline alongside prairie dogs, including the swift fox, the burrowing owl, the golden eagle, the ferruginous hawk, the mountain plover, and the black-footed ferret (swift fox, *Vulpes velox,* Knowles and Knowles 1994; burrowing owl, *Athene cunicularia,* Knowles and Knowles 1994; Desmond et al. 2000; golden eagle, *Aquila chrysaetos,* Cully 1991; ferruginous hawk, *Buteo regalis,* Knowles and Knowles 1994; Miller et al. 1994; mountain plover, *Charadrius montanus,* Knowles and Knowles 1994; black-footed ferret, *Mustela nigripes,* Clark 1989). The extinction of prairie dogs would cause the disruption of the grassland ecosystem and would likely cause the extinction of other species as well.

From an ethical standpoint, prairie dogs have some qualities that are similar to humans. They have a sophisticated communication system and a complex social system (see chapters 3 and 4). They have much to teach us about the ability of animals to process information and make decisions about events in the world around them. As we move away from the paradigm that animals are unfeeling, unthinking automatons that run on instinct, we see that they are actually complex creatures that share some of the characteristics of humans.

When examined as an isolated incident, the decline of prairie dogs may be uncomfortable, sad, and unethical to some, yet quite easily overlooked by others. However, it should not be overlooked by anyone because the decline of prairie dogs is a universal tragedy. Viewed from a wider perspective, the prairie dog decline is part of a larger web of degradation that human society is imposing on the natural world. This degradation is perpetuated by a complicated set of social, economic, political, and historic forces, and its outcome is potentially very dangerous to the physical, emotional, and ethical health of humans and to the other species that share the earth with us. At present, human society only glimpses signs of this degradation like a whale surfacing occasionally from the ocean. Each time we encounter another case of species in decline we see the whale's slick back curving up. But, as a society, we fail to recognize the bulk of the entire beast beneath the waters and the speed at which it is progressing. With the current prairie dog plight the figurative whale has surfaced again and we have the opportunity to examine our full relationship with the earth and all its complexity and depth. Thus, by understanding the interacting forces surrounding prairie dog decline, we start to unravel clues that help us understand the underlying problems inherent in the way we

humans currently relate to the natural world. Understanding these interactions is the first step in shifting us toward a more sustainable future.

Biotic Impoverishment: Prairie Dogs, Biodiversity, and the Big Picture

Just as the health of many species depends on human action, so the well-
being of human societies is linked to the fate of non-human species. Today,
more than ever, we are starting to understand and feel the consequences—
both emotionally and physically—of the extensive loss of biodiversity,
of which prairie dogs are but one part. Although most people think of
biodiversity as the number of different species in the world, it is actually
much more encompassing and much more important than that. Biodiversity comprises a hierarchical structure of the diversity of life from the
level of individual genes, to species, to whole ecosystems and communities (Congressional Biodiversity Act, HR1268 (1990)). Life has evolved this
diversity over billions of years and cannot exist without it. It is the fabric
that creates life processes and holds earth's life support systems together
(Loreau 2000). Biodiversity is structured through an interlinked system; the
loss of biodiversity on one level causes ripples of change through all levels
and environments. Thus, while this book focuses on the loss of biodiversity
at the level of a single species, we will also examine the wider consequences
of the loss of prairie dog species in terms of the resulting decline in genetic
and ecological diversity.

Species are one of the fundamental units that keep ecosystems functioning.
A functioning ecosystem is essential to maintaining a clean environment and
equable climate regimes, and preventing disease outbreaks; a loss of biodiversity therefore impacts the sustainability of human societies. The earth provides
us with the air we need and the energy that flows through ecosystems. Forests
provide us with a delicate balance of oxygen and carbon dioxide. The species
of plants and animals in forests, grasslands, oceans, rivers, shrublands, and
deserts keep energy, water, and nutrients cycling through ecosystems so that
we can exist. The well-being of most species today is closely linked with past
human actions as well as the course of human action into the future. The
rate of species extinctions today is more than 1,000 times higher than background, or non-human-caused, extinction levels (Nott et al. 1995). Prairie

dogs are in the company of almost 20% of all mammal species on earth today that are currently under threat of extinction (Chapin et al. 2000).

Human emotional well-being is also linked to the sustenance of the world's natural species and features. Some researchers in biology, psychology, and sociobiology have shown that we are dependent on nature in many ways. Being around natural systems and species enhances general happiness, fosters social connections, increases rates of healing, and possibly even determines whether we reach our full human development potential (Kaplan and Kaplan 1989; Kellert and Wilson 1993; Kahn and Kellert 2002; Kahn 2001). E. O. Wilson's Biophilia hypothesis (Wilson 1993) asserts that humans have an innate affiliation with nature that has fostered our continued survival as a species throughout time. Even the richness of our cultures depends on the existence of varied plants and animals with which to interact.

The good news is that conservation efforts worldwide have boomed in the last 20 years and are launching a formidable response to this increasing biotic impoverishment. In addition, new multidisciplinary scientific fields such as conservation biology, sustainability science, and restoration ecology have formed since the mid-1980s and have fostered many ways of trying to meet this challenge. However, we are still far from being able to deal with the complex set of forces that lead human societies to degrade the natural world. Even the issue of the conservation of one species, such as the prairie dog, is daunting because it is tied to major economic, social, and political forces occurring over many spatial and temporal scales.

This places higher demands on those of us who want to learn about prairie dogs or other species. We must not look only at biological and physical aspects of prairie dogs; we must also venture into a multi-disciplinary journey that touches on a variety of interactions that prairie dogs have with the greater ecosystem in which they live and with the greater human society with which they are forced to interact.

The story of the prairie dog is important because it is not just the story of one species in isolation. Drastic changes, including extensive habitat loss and severe population declines over the past hundred years, are not events unique to prairie dogs. They are themes that carry through the stories of many other species, ecosystems, and global systems—the many tiers of biological diversity. In these stories we see a recurring pattern occurring in the way that humans relate to the natural world. This pattern creates widespread biotic impoverishment of which the full consequences are not yet known. We also do not yet understand the full extent of the causes associated with this pattern. But the more we explore individual cases in detail, and the more we are able to tease out the underlying causes of biotic impoverishment, the more we will be able to uncover and create avenues of positive change. The basic challenges in information gathering and synthesis, of acting in the face of uncertainty, and of dealing with conflicts between current economic, social, political, and natural systems are the same for all species. For this reason, the themes discussed in this book, while specifically tailored to the prairie dog, can be applied to many other conservation situations. Our goal is that by looking at the case of prairie dogs in detail we will not only be able to increase our effectiveness in pulling prairie dogs back from the brink of extinction, but we will also be able to use the lessons that prairie dogs teach us in gaining an understanding of the larger conservation challenge that we are currently facing.

2

The Life of Prairie Dogs

In our story of prairie dogs, we first need to know something about who they are and where they live. Prairie dogs are social ground squirrels that live in colonies (often called towns) on the grasslands of North America. They are found in the central portion of the United States along a broad swath that extends from Canada to Texas and from Arizona to the Dakotas. As the grasslands have fallen to farmland, development, parking lots, towns, cities, and roads so have the fortunes of the prairie dogs declined. At the beginning of the 20th century, there were an estimated 5 billion prairie dogs in North America. Now, a little more than 100 years later, there are about 1%–2% of their former numbers left and the numbers are rapidly dwindling.

The Five Species of Prairie Dogs

Prairie dogs are formally classified as rodents within the squirrel family, Sciuridae. There are five prairie dog species (Hall and Kelson 1959) (Figures 2.1–2.4). The black-tailed prairie dogs *(Cynomys ludovicianus)* have the broadest distribution (Figure 2.5), from southern Canada in the north to Chihuahua, Mexico, in the south; Arizona (historically, although they are currently extinct in this state), Colorado, and Montana in the west; and the Dakotas and Nebraska in the east. The Gunnison's prairie dogs *(Cynomys gunnisoni)* have the next largest distribution, and are found in northern Arizona, northern New Mexico, southwestern Colorado, and southeastern Utah. The white-tailed prairie dogs *(Cynomys leucurus)* have a more limited distribution in western Colorado, southwestern Wyoming, and just barely into southern Montana. The Utah prairie dogs *(Cynomys parvidens)* are found in only a

Figure 2.1. Black-tailed prairie dog. (Courtesy of Dean Biggins.)

few locations in central and southern Utah. Finally, the Mexican prairie dogs *(Cynomys mexicanus)* are found in a few locations in central Mexico. Of these five species the Mexican prairie dog is listed as endangered by international CITES standards (CITES stands for the Convention on International Trade in Endangered Species of Wild Fauna and Flora, an international agreement drawn up in 1973, now with 172 member governments, to ensure that endangered species are not jeopardized by international trade and under the ESA since 1970); the Utah prairie dog is listed as threatened by the U.S. Fish and Wildlife Service; and the black-tailed prairie dog was listed until recently as a candidate species, which means that the U.S. Fish and Wildlife Service found

Figure 2.2. Gunnison's prairie dog. (Photo by C. N. Slobodchikoff.)

sufficient reasons for listing the species as either threatened or endangered but there was no money available for the actual listing. All of the species of prairie dogs look fairly similar (Hoogland 1995) with brown or tan fur, a blunt nose, small mouse-like ears, and a short stubby tail. The color of the tail distinguishes the two major groups of prairie dogs. Two species, the black-tailed and the Mexican, have black at the tip of their tails (and are corre-spondingly classified in the genus *Cynomys*, subgenus *Cynomys*), while three species, the Gunnison's, white-tailed, and Utah, have white at the tip of their tails (and are classified in the genus *Cynomys*, subgenus *Leucocrossuromys*). The black-tailed group tends to be somewhat larger, while the white-tailed

Figure 2.3. White-tailed prairie dog. (Courtesy of Dean Biggins.)

prairie dogs tend to be somewhat smaller. Adult prairie dogs are between 25 and 40 centimeters long and weigh between 300 and 900 grams in the spring and 500 and 2,000 grams in the fall.

Other than the color of the tail and the geographical areas they occupy, there are a few minor differences (Hoogland 1995). Black-tailed and Mexican prairie dogs have longer tails (in the range of 60–100 millimeters for black-tailed and 90–110 for Mexican), while the white-tailed group has shorter tails (in the range of 30–65 millimeters). The females of the black-tailed group have 8 teats, while the females of the white-tailed group have 10 teats. The white-tailed and the Utah prairie dogs have a black spot above each eye,

Figure 2.4 Utah prairie dog. (Courtesy of Dean Biggins.)

while the other species do not have a black spot. The two species in the black-tailed group do not hibernate in the winter and are active year-round except in inclement weather when they go into short-term torpor (Harlow and Menkens 1986; Harlow and Frank 2001; Lehmer and Van Horne 2001). The three species in the white-tailed group hibernate inside burrows during the winter where they curl up into a C-shape and their body temperature drops to the point that their muscles lock so that the animals cannot be uncurled until their body temperature rises (Rayor et al.1987; Bakko and Nahorniak 1986). All of the species of prairie dogs, except the Gunnison's, have 50 chromosomes. The Gunnison's prairie dog has 40 chromosomes (Nadler et al. 1971).

Evolutionary History of Prairie Dogs

Evidence from DNA sequences confirms that, in an evolutionary sense, the black-tailed and the Mexican prairie dogs are closely related to each other, while in the white-tailed group, the white-tailed and the Gunnison's prairie dogs are close relatives, with the Utah prairie dog being a slightly more distant relative of the other two species. Prairie dogs are more distantly related to

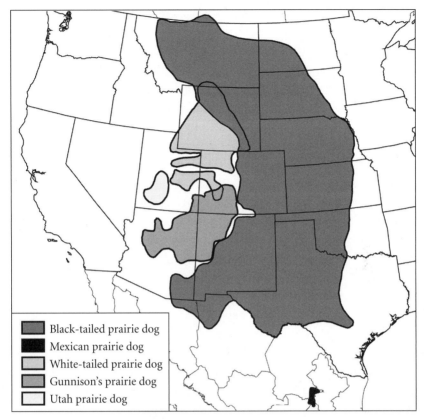

Figure 2.5. Historical range of the five species of prairie dogs. (From Wagner et al. 2006, with permission from Elsevier.)

the marmots *(Marmota)*, a group that includes the woodchuck *(Marmota marmota)* and the yellow-bellied marmot *(Marmota flaviventris)* (Herron et al. 2004).

As a group, prairie dogs apparently date back more than 2 million years (Goodwin 1995). Fossil prairie dogs have been found from the late Pliocene period (late Blancan period, 2.5–1.8 million years BP) in the Great Plains of North America. At present, it has not been possible to determine if these are members of the black-tailed or the white-tailed group. By the early Pleistocene (early Irvingtonian period, 1.8 million–750,000 years BP), representative fossils of both the black-tailed and white-tailed groups appeared on the central Great Plains. During the middle

Irvingtonian period (750,000–500,000 years BP), Gunnison's prairie dogs *(Cynomys gunnisoni)* appeared in the fossil record in southern Colorado, the oldest known prairie dog species that is still present today. In the late Irvingtonian period (500,000–300,000 years BP), an extinct white-tailed species, *Cynomys niobrarius,* was distributed from southern Canada to central Kansas, and an extinct black-tailed species was found from southern Nebraska to northern Texas. By the Sangamon period (125,000–75,000 years BP) *C. niobrarius* was found from eastern Idaho to Saskatchewan to southern Kansas. The present-day black-tailed prairie dog *(Cynomys ludovicianus)* appeared during the Wisconsin period (75,000–10,000 years BP) and was found from central Nebraska in the north to southwestern New Mexico and possibly as far south as southeastern Chihuahua, Mexico. So far no fossils of the Mexican, the white-tailed, or the Utah prairie dogs have been found (Goodwin 1995).

The fossil and genetic evidence suggests a possible scenario for the evolution of the present-day species of prairie dogs. The common ancestor of the prairie dogs was probably a species much like the Gunnison's prairie dog *(Cynomys gunnisoni)* originating somewhere on the Great Plains of North America (Pizzimenti 1975; Goodwin 1995). The Gunnison's prairie dog probably evolved in the southern Rocky Mountains (Pizzimenti 1975), and the two more northern white-tailed species, the white-tailed prairie dog and the Utah prairie dog, evolved relatively recently from the extinct *Cynomys niobrarius* that had a more northern distribution than the Gunnison's prairie dog (Goodwin 1995). The close evolutionary relationship between the white-tailed and the Utah prairie dogs is supported by immunological evidence, which suggests that the two species are antigenically very similar, with the Gunnison's prairie dog a more distant relative on the basis of immune system antigens (McCullough et al. 1987). However, the DNA sequence data suggests that the Gunnison's and the white-tailed prairie dogs are closer relatives than the Utah (Herron et al. 2004). The black-tailed prairie dog probably evolved from a black-tailed group ancestor, and the Mexican prairie dog probably evolved relatively recently from the black-tailed prairie dog (Goodwin 1995). Genetic evidence based on allozymes suggests that Mexican prairie dogs diverged from black-tailed prairie dogs approximately 42,180 years ago (McCullough and Chesser 1987).

Feeding Habits

The feeding habits of the different prairie dog species are fairly similar. All are herbivores who eat plant material that includes leaves, stems, roots, and seeds. Grasses make up a main component of the diet; herbaceous plants such as forbs and shrubs make up another major component. In a colony of black-tailed prairie dogs at Buffalo Gap National Grassland, South Dakota, grasses made up an average of 70.9% of the annual diet of the prairie dogs (Fagerstone et al. 1981). During the summer months, grasses made up 72.9% of the plant species growing at the site and 74.2% of the ground cover, so in the summer prairie dogs were feeding on the most dominant plants in roughly the same proportion as the occurrence of the plants. Wheatgrass (*Agropyron* spp.), buffalograss *(Buchloe dactyloides)*, and blue grama *(Bouteloua gracilis)* made up an average of 55.8% of the diet. In a study of black-tailed prairie dogs near Wall, South Dakota, grasses made up 87% of the diet, and the prairie dogs fed selectively on certain grasses and grass-like plants such as ring muhly *(Muhlenbergia torreyi)*, green needlegrass *(Stipa viridula)*, and sand dropseed *(Sporobolus cryptandrus)* (Uresk 1984). There was only a 25% similarity in the composition of the plants growing at the site and the composition of the plants in the prairie dog diets, indicating that the prairie dogs had distinct preferences toward certain plant species. Other studies have also found that grasses are dominant food items for black-tailed prairie dogs (Kelso 1939; Koford 1958; Smith 1967; Tileston and Lechleitner 1966; Summers and Linder 1978). A study of the feeding habits of Mexican prairie dogs *(Cynomys mexicanus)* at a colony near San Luis Potosi, Mexico, showed that the animals most frequently ate one grass *(Muhlenbergia repens)* and one forb (*Halimolobos* spp.)(Mellink and Madrigal 1993). A study of the stomach contents of 169 white-tailed prairie dogs, collected from several locations in Montana and Wyoming, showed that an average of 51% of the contents were forbs and shrubs such as saltbush (*Atriplex* spp.) and Russian thistle *(Salsola pestifer)*, and an average of 28% of the contents were grasses, primarily wheatgrass (*Agropyron* spp.) (Kelso 1939). Gunnison's prairie dogs near Flagstaff, Arizona, ate a mixture of grasses and forbs but preferred blue grama *(Bouteloua gracilis)*. They ate this grass in higher proportions than it occurred within the general habitat (Shalaway and Slobodchikoff, 1988). Arthropods, such as insects, generally are not eaten by prairie dogs; low percentages (1%–2%)

show up in fecal samples or stomach contents, perhaps as a result of accidentally eating the insects while munching on forbs or grasses (Fagerstone et al. 1981; Shalaway and Slobodchikoff 1988).

The feeding preferences of prairie dogs shift seasonally. At Buffalo Gap, South Dakota, black-tailed prairie dogs had between 82% and 92% grass in their diets in December, May, and July (89.4% in December, 92.0% in May, 82.6% in July). In other months, the grass in their diets varied between 39% and 66% (56.1% in January, 39.0% in February, 66.4% in September) (Fagerstone et al., 1981). At Wall, South Dakota, the composition of grasses in the diet varied between a low of 74.3% in September and a high of 94.8% in June (Uresk 1984). Similarly, the diet of Mexican prairie dogs near San Luis Potosi changed seasonally, with grasses making up the bulk of the diet in April and May, and forbs the majority of the diet in June, July, and August (Mellink and Madrigal 1993). For white-tailed prairie dogs, shrubs and forbs made up between 72% and 100% of the diet in December through March, and in July grasses made up 62% of the food eaten by the prairie dogs (Kelso 1939). Gunnison's prairie dogs in the vicinity of Flagstaff, Arizona, fed on seeds early in the season in the spring, after the adults emerged from hibernation, then switched to grasses and forbs as the green vegetation started to grow in the summer, and then shifted back to seeds as the green vegetation died back in the fall (Shalaway and Slobodchikoff 1988).

A generalization that can be made about the food habits of prairie dogs is that they are variable. Grasses are preferred food items, but the species of grasses eaten varies from one site to another. This may be because the vegetation growing on different colonies is highly variable—some colonies live in places where there is mostly grass, other colonies live in places where there are mostly forbs. The species composition of plants changes from one colony to another. Seeds are also eaten where available (King 1955; Fagerstone et al. 1981; Uresk 1984; Shalaway and Slobodchikoff 1988).

Groups, Territories, and Burrow Systems

Within a colony, prairie dogs live in groups that defend territories. Each group, sometimes called a coterie (King 1955; Hoogland 1995), occupies an area that includes one or more burrow openings, an underground burrow system, and the food resources growing within the territory. Territorial boundaries are usually defended by all the members of the territorial group. Group size within a territory can vary from a single animal to groups that contain several males and several females (see chapter 3).

The burrow system represents an important resource for the prairie dogs on their territories. Because burrows are energetically expensive to dig, prairie dogs prefer to use burrows that are already constructed, and can inhabit burrow systems that might have persisted for tens or hundreds of years (King 1984). As a result, additions or repairs are often made to existing burrows (Longhurst 1944). Prairie dogs spend a considerable portion of their life underground seeking refuge from predators, inclement weather, and seasonal climate changes. White-tailed prairie dogs reportedly spend close to two-thirds of their life in burrows (Clark et al. 1971). Black-tailed prairie dogs live inside their burrows for more than half of the time that they are alive (Hoogland 1995).

Burrow systems can be simple tubes running down into the ground, or they can be complexes of tunnels with multiple entrances and exits (Figure 2.6). Tunnels are usually 5–10 centimeters in diameter, going down to depths of 2–5 meters, and can be 5–35 meters long. Information on burrows is usually obtained through excavation in soils that sometimes have the consistency of concrete therefore; our knowledge of burrow architecture is relatively sparse (Flath and Paulick 1979). An excavation of a white-tailed burrow system (Burns et al. 1989) found that shortly after the entrance of one mound there was a "turning bay" where an animal could turn in the burrow, or let another animal go past (Scheffer 1937). Following the turning bay was a series of intersecting tunnels. The discovery of fresh and semi-fresh plant matter distributed at various points in the burrow network (Burns et al. 1989) suggests that prairie dogs may cache food. Burns et al. (1989) describe a steep slope to both main burrow entrances, a common feature of prairie dog burrows. These steep inclines may aid in reducing the flooding potential of the burrow. The white-tailed prairie dog burrow system extended for a total length of 29.3 meters, a depth of 2 meters, and contained five side branches. In contrast, Gunnison's prairie dog burrows are smaller than white-tailed prairie dog burrows. Longhurst (1944) reported a correlation between the size of a burrow and the steepness of the slope on which it was located. He suggested that on steeper slopes, burrows have a higher run-off of excess water and therefore need not be as large as those on the plains. Gunnison's

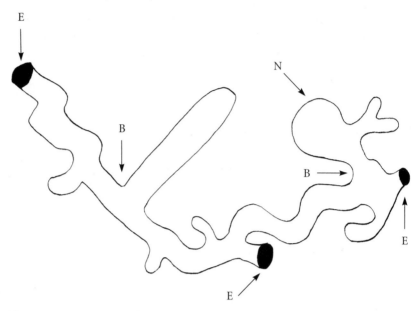

Figure 2.6. Burrow system of Gunnison's prairie dog. There are three entrances (E), a nest
        chamber (N), and two major right-angle bends (B). This burrow system was
        excavated in the vicinity of Flagstaff, Arizona.

prairie dog burrows range in depth from approximately 1–2 meters and have
nesting areas for hibernating prairie dogs (Smith 1982; Verdolin et al. 2008).
A burrow system of captive Utah prairie dogs excavated at the National Zoo
revealed no basic differences from other prairie dog burrows (Egoscue and
Frank 1984). Black-tailed prairie dog burrows range from the simple to the
complex. A complex burrow system described by Hoogland (1995) had two
entrances, two major branches from the main tunnel, a nest with juveniles at
a depth of 2 meters, a tunnel with a large chamber at a depth of 1 meter, and a
chamber or turning-bay at a depth of 0.8 meters, which branched away from
one of the main tunnels.

    Surface openings of the burrows can be either in the form of mounds or
in the form of holes that are relatively level with the ground. Black-tailed
and Mexican prairie dogs have some mounds (called rim craters) that look
like miniature volcanos, and other mounds that are more rounded (called
dome craters) (Hoogland 1995). The white-tailed group often has mounds
with multiple entrances within the mound, while the black-tailed group

rarely has more than two openings within a mound. The rim craters of black-tailed prairie dogs can create airflow within the tunnels when the wind blows past the rim crater pulling air out of the tunnels (Vogel et al. 1973). The more rounded mounds of the white-tail prairie dogs may have a similar function.

Burrows offer prairie dogs protection from predators and a place to raise their young, but they also offer a more stable microclimate than the outside environment. The humidity is generally higher inside the burrows, and the temperature fluctuations are less extreme, giving the animals a cool place in the summer and a warmer place in the winter. In the winter, burrow temperatures can be 10°C warmer than the outside ambient temperatures. A study of burrow temperatures of Gunnison's prairie dogs found that in January surface temperatures fluctuated between a high of 13.3°C and a low of -7.2°C; burrow temperatures 2 meters below the surface fluctuated between 4.2°C and 2.6°C. While both surface and burrow temperatures rose from January through July, the burrow temperatures had less fluctuation on a daily basis (Smith 1982).

A disadvantage of the stable microclimate offered by burrows is that it provides a sheltered place for parasites such as fleas to breed. Prairie dogs have a number of different ectoparasites, primarily fleas and ticks. Tick species found living on prairie dogs include *Ixodes kingi, Atricholaelaps glasgowi* (Hoogland 1995), and *Ornithodoros parkeri* (Gage et al. 2001), which can inhabit burrows. The life history of fleas particularly lends itself to burrow life. Fleas lay their eggs either directly on their host or on any material that is in the general vicinity of the host. As the flea larvae hatch, they begin feeding on any organic debris found in their vicinity, including dried blood defecated by the adult fleas. The fleas then metamorphose into pupae that can live for up to 50 weeks without the presence of a host, and as adults fleas can live up to one year. Burrows provide a sheltered environment for the fleas, particularly the larvae, and flea populations inside burrows can become very large (Seery et al. 2003). Although some 20 species of fleas have been recorded from prairie dogs burrows three species are particularly important because they can easily transmit plague (see Prairie Dogs and Plague, pages 29–31) These three species are *Oropsylla hirsuta, O. labis,* and *O. tuberculata* (Anderson and Williams 1997; Cully et al. 1997).

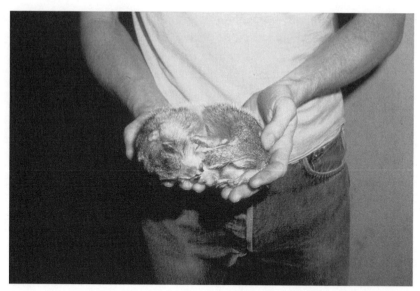

Figure 2.7. Gunnison's prairie dog in hibernating position. During hibernation, the muscles lock and the prairie dog cannot be uncurled from this posture. (Photo by C. N. Slobodchikoff.)

Hibernation

Perhaps as a way of avoiding bad weather the prairie dogs in the white-tailed group hibernate in the winter and thus avoid negative net energy balance. Adult animals start to disappear belowground in August or September, and reappear again in late February or early March (Bakko and Nahorniak 1986). Once belowground they enter hibernation. This is not the type of hibernation where the animal goes to sleep and then wakes up months later. In prairie dog hibernation, which has been called torpor, the animals curl up and allow their body temperature to drop from above 29°C to about 5°C–15°C; the muscles lock and the body remains in that condition for an average of eight days (Smith 1982; Bakko and Nahorniak 1986; Rayor et al. 1987) (Figure 2.7). They wake up and raise their body temperature to above 29°C and are active in their burrows for an average of 21 hours before lowering their body temperature again to repeat the cycle (Bakko and Nahorniak 1986). White-tailed prairie dogs spend an average of 86% of their time belowground during winter in the condition of torpor, with a low-body temperature (Bakko and Nahorniak 1986), and Gunnison's and Utah prairie dogs very likely spend a comparable amount of time. Utah prairie dogs spend intermittent times in torpor at the beginning and the end of the winter season (Lehmer and Biggins 2005). While in torpor, they live on stored body fat. Before entering into torpor for the winter season the white-tailed prairie dogs' body weight is about 27% fat (Harlow and Buskirk 1996).

Unlike the white-tailed group, which always goes into hibernation in the fall, the black-tails go into torpor or limited hibernation sometimes, but not always. (When an animal sometimes goes into hibernation but not always, it is known as a facultative hibernator, and when an animal always goes into hibernation, it is known as an obligate or spontaneous hibernator.) For a long time, people believed that black-tails did not hibernate at all, but remained active throughout the year. Recently, several studies have shown that black-tailed prairie dogs in both the laboratory and in the field will go into torpor for short periods of time (Harlow and Menkens 1986; Harlow and Frank 2001; Lehmer and Van Horne 2001; Lehmer et al. 2001; Lehmer and Biggins 2005). Black-tailed prairie dogs in the laboratory go into torpor primarily when they are cold, at a temperature of 6°C, and when they have not been able to eat. Five black-tailed prairie dogs that were monitored in the wild all went into torpor for periods of time averaging 141 hours (about six days), and the animals went into torpor between five and nine times during a four-month period from December through March regardless of ambient temperatures and food conditions (Lehmer et al. 2001). Elevation also seems to play a role. Black-tailed prairie dogs from higher elevation colonies have been reported going into torpor earlier, and for longer periods of time, than prairie dogs from lower elevation colonies (Lehmer and Biggins 2005). However, in another study of black-tailed prairie dogs, wild prairie dogs that were tracked over the space of a year never went into torpor, maintaining a body temperature that varied between 39.3°C in the daytime in August to 33.5°C at night in January (Bakko et al. 1988). A combination of food deprivation and a diet of plant oils rich in polyunsaturated fatty acids seems to increases the frequency and length of time of torpor of both white-tailed and black-tailed prairie dogs (Harlow and Frank 2001).

Reproduction

Another function of burrows is to facilitate reproduction. Most of the species of prairie dogs mate underground (the white-tailed prairie dog is said to mate above-ground), and the pups are born in nest chambers underground (Hoogland 1995). Prairie dogs are rodents, and a popular conception of rodents is that they are constantly breeding; this is not the case with prairie dogs. These animals breed only once a year and have relatively few pups. The females are receptive for less than a day, which is the only opportunity for males to mate with them (Hoogland 2001). During that time the female can mate with multiple males, each potentially fathering one or more pups as a result of that mating (Hoogland 1995; Travis et al. 1996).

Mating is in the winter or spring months, depending on the species. Like some other mammal species, the testes of the males descend and develop sperm only during the breeding season, after which they are pulled up into the body and are not capable of producing sperm (Anthony 1953). Black-tailed prairie dog males start to produce functional sperm during December, and sperm production can continue through April (Anthony 1953). Mating takes place between February and April (Hoogland 1995). Gunnison's, Utah, and white-tailed prairie dogs have a shorter breeding season beginning in mid-March and ending in early April (Bakko and Brown 1967; Knowles 1987; Hoogland 1999).

Litter sizes of prairie dogs tend to be small. Average litter sizes are 3.08 pups per female for black-tailed, 3.77 pups for Gunnison's, and 3.88 pups for Utah prairie dogs (Hoogland 2001). Litter sizes for white-tails and Mexican prairie dogs are probably comparable. Stockard (1929) reported that female white-tails near Laramie, Wyoming, had an average of 5.48 embryos in their uterus, and Bakko and Brown (1967) reported that female white-tails from southeastern Wyoming had an average of 5.64 embryos. Not all of those would be expected to mature to the point of being born. The average length of gestation for black-tailed prairie dogs is 34.6 days, and the average gestation for Gunnison's prairie dogs is 29.3 days (Hoogland 1995; Hoogland 1997). Lactation by females is estimated at 41.3 days for black-tails and 38.6 days for Gunnison's (Hoogland 1995; Hoogland 1997).

Not all individuals successfully copulate or conceive during the breeding season. Should a female conceive and successfully wean an entire litter ranging from 2–6 pups, survivorship of her pups past the first year is low. Survivorship differs very little between male and female pups with black-tail, Gunnison's, and Utah prairie dogs. All male pups have less than 50% survivorship to the age of one year; for females the probability of living past one year ranges from 50%–54% (Hoogland 2001). It is unclear what factors are responsible for the low survival rate of pups, but it is likely that competition, predation, and over-winter hibernation contribute to such high mortality rates.

Just as with the pups, mortality of adult prairie dogs is also high. Hoogland (1995) reports a summary of survivorship of black-tailed prairie dogs at Wind Cave National Park over a 15-year period showing that in the first year of their life, 53.2% of the males and 45.7% of the females died. Out of 587 males observed at first emergence from their burrows after their birth, only eight lived to the age of five years, and none lived to age six. Females lived slightly longer than males. Out of 523 females observed at first emergence only four lived to age seven, one lived to age eight, and none lived to age nine. This means that most adults who survive their first year have about two reproductive seasons to produce offspring, after which their chances of reproduction and survival decline precipitously.

Vision and Hearing of Prairie Dogs

To detect predators and communicate information about the predators to others in the colony, prairie dogs need good vision and hearing. Prairie dog retinas have both rods and cones (Jacobs and Pulliam 1973; Jacobs 1981). The rods are sensitive to light in the 500 nanometer range (this is in the range of blue light, but we would see this as gray or black and white, with bluer objects appearing brighter). The cones come in two types: one has photopigments that are sensitive to a peak of about 440 nanometers (blue to our eyes) and the other has photopigments that are sensitive to a peak of 525 nanometers (yellow-green). This means that prairie dogs have color vision, but the vision is dichromatic, allowing the prairie dogs to see well in the blue-green-yellow range, but not as well in the red range (Figure 2.8). In contrast, humans have trichromatic vision, with three types of cones, each of which is sensitive to either a peak of 420 nanometers (blue), 534 nanometers (yellow-green), or 564 nanometers (yellow-red).

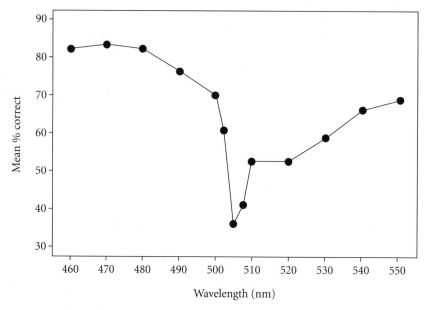

Figure 2.8. Visual discrimination by black-tailed prairie dogs of different wavelengths of
light. In a narrow zone around 505 nm, the prairie dogs do not discriminate
between those wavelengths and achromatic light of equal brightness. All other
wavelengths in the figure are correctly discriminated from achromatic light.
Chance performance in this discrimination is 33.3 percent. (From Jacobs and
Pulliam 1973, with permission.)

Prairie dogs can hear sounds in much the same range as humans (Heffner
et al. 1994) (Figure 2.9). A young, healthy human can hear sounds in the
range of 20–20,000 Hz (where Hertz is a unit that represents the vibration
of the sound wave, the number of cycles per second). Within this range,
humans can hear sounds best when the sounds are from 500 to 4,000 Hz.
Black-tailed prairie dogs can hear sounds from 29–26,000 Hz, while white-
tailed prairie dogs can hear sounds from 44–26,000 Hz at normal levels of
loudness (60 decibels). When the sound is very loud (90 decibels), black-
tailed prairie dogs can hear sounds of 4 Hz or very low-frequency sound.
Much like humans, prairie dogs can hear sounds best when they are in the
range of 500–4,000 Hz. In contrast, another rodent, the Norway rat, cannot
detect sound frequencies below 250 Hz, and hears sound into the ultrasonic
range, up to 64,000 Hz (24).

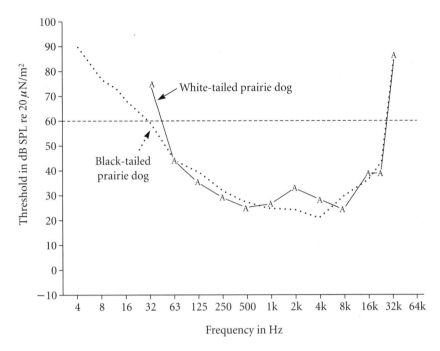

Figure 2.9. Audiogram of white-tailed and black-tailed prairie dogs showing sensitivity to different sound frequencies. Solid line represents the response of white-tailed prairie dogs, and dotted line represents the response of black-tailed prairie dogs. Both species are relatively insensitive to very low frequencies and very high frequencies, with the maximum sensitivity within the 500 Hz to 8,000 Hz range. (From Heffner et al. 1994, with permission from Elsevier.)

Predation and Vigilance

A number of predators eat prairie dogs (see also chapter 6). These include both terrestrial and avian predators. Raptors that prey on prairie dogs include red-tailed hawks *(Buteo jamaicensis)*, Swainson's hawks *(Buteo swainsoni)*, golden eagles *(Aquila chrysaetos)*, ferruginous hawks *(Buteo regalis)*, rough-legged hawks *(Buteo lagopus)*, harriers *(Circus cyaneus)* (Slobodchikoff et al. 1991), and goshawks *(Falco mexicanus)*. Prairie dogs are also a staple in the diet of many terrestrial predators, including coyotes *(Canis latrans)*, gray foxes *(Urocyon cinereoargenteus)*, swift foxes *(Vulpes velox)*, badgers *(Taxidea taxus)*, long-tailed weasels *(Mustela frenata)*, black-footed ferrets *(Mustela nigripes),* and rattlesnakes *(Crotalus atrox; Crotalus viridis viridus).*

Prairie dogs are the principal food of the endangered black-footed ferret. In areas of South Dakota, nearly 50% of the diet of swift foxes can be composed of prairie dogs (Uresk and Sharps 1986). Also, badgers can rely on prairie dogs for over 50% of their dietary needs (Goodrich and Busirk 1998). Among the raptors, the dependence is more variable and fluctuates with the density of prairie dogs. Golden eagle diets in New Mexico can include 18% prairie dogs while in areas where ferruginous hawks nest the bulk of their diet may be composed of prairie dogs (Cully 1988; Cully 1991).

Hunting prairie dogs is not an easy task, however. Even though many predators attack and eat prairie dogs, the social system and the alarm call system keep the predators from being too successful. In a study of predation and vigilance in Gunnison's prairie dog colonies near Flagstaff, Arizona (Verdolin and Slobodchikoff 2002), 142 predation events (where a predator appeared to be hunting on or near the colony) were observed, of which 16.2% resulted in an attempt to capture a prairie dog. In total, four kills were observed, such that 17.4% of all attempts were successful. Aerial predators were responsible for 74% of all attacks undertaken, and three of the four kills observed, resulting in a 17.6% success rate. Though terrestrial predators occurred less frequently, attacked less often, and made a single kill over the study period, the calculated success rate is similar to that of aerial predators at 16.7%.

By living in groups, animals may increase the likelihood of their survival. The more individuals that are present the greater is the probability of detecting a predator sooner (Hoogland 1981; Clutton-Brock et al. 1999). Larger groups may afford individuals an opportunity to spend less time being vigilant and more time feeding (Berger 1978; Caraco 1979; Lima 1995). Through dilution, the probability of attack for any individual may decline as group size increases (Berger 1978; Clutton-Brock et al.1999). Prairie dogs spend a considerable amount of time being vigilant, either scanning for predators or standing upright in an alert posture. In the study of vigilance in Gunnison's prairie dogs there was no significant difference in the amount of time the animals spent being vigilant on smaller colonies or larger ones (Verdolin and Slobodchikoff 2002). However, a study of black-tailed prairie dogs experimentally manipulated the group size of different groups and found that groups that were experimentally made smaller tended to forage more alertly and in a less risky manner (Kildaw 1995). Many predators eat prairie dogs, but they are not all that easy to catch.

Prairie Dogs and Plague

A major threat to the survival of the prairie dog comes from bubonic plague. Plague is a disease caused by a bacterium *(Yersinia pestis)*. In animals it can occur in a sylvatic cycle, which means that it is present in low levels in a variety of mammal species in the wild (Keeling and Gilligan 2000). During this cycle, it is transmitted by fleas from one host to another. In historic times, plague would go from a sylvatic cycle to an urban cycle when it would infect rats that were living in human houses in cities and towns, and the fleas living on the rats would transmit the disease to humans. In humans who contracted the disease, the lymph nodes in the vicinity of the flea bite would swell and turn a bluish-black—these swollen lymph nodes are known as buboes, hence the name bubonic plague. Among both animals and humans, plague can go into a pneumonic cycle when it infects the lungs, and in the pneumonic cycle the plague bacilli can be transmitted from animal to animal, animal to human, or human to human by coughing or sneezing. From about 500–1700 AD, plague was responsible for killing large numbers of people in Asia and Europe, and changed the social fabric of medieval society by killing perhaps two-thirds of the population of Europe. Currently, the incidence of plague in humans in the United States is low (Barnes 1982). From 1925 to 1946, the average number of cases was two per year. Since then, this average has increased to 16 per year. Fortunately, plague can be treated successfully with common antibiotics (Pollitzer 1954; White et al. 1980).

Some species of fleas are better at transmitting plague than others. In some species, the plague bacilli form a plug in the digestive tract of the flea (in a part of the gut known as the proventriculus), plugging up the tract and making it difficult for the flea to feed. As the flea tries to free the plug, it regurgitates the blood that it has taken in, and in the process injects numerous plague bacterial cells into the bloodstream of its host. Because the flea cannot feed, it tries to get a blood meal by jumping off one host and onto others and can inject plague bacilli into numerous hosts. The bacterial plugs can kill infected fleas in some cases. In other instances, the infected fleas can live up to a year with plague in their gut after the prairie dogs have disappeared (Lechleitner et al. 1968).

Plague is not native to North America. It came to the United States in 1899–1900, probably from Asia, showing up among rats in San Francisco (Barnes 1982). Since that time it has been moving eastward, and currently appears to be expanding past the 100th meridian of the United States. It has not been documented in Mexico (Trevino-Villarreal et al. 1998), although there are some indications that it might have arrived there. More than 200 species of mammals from 95 genera can either be hosts or can become infected with plague, testing positive for the antibodies. In the United States these species include domestic dogs, domestic cats, rock squirrels, ground squirrels, skunks, raccoons, bobcats, coyotes, badgers, foxes, and prairie dogs (Poland and Barnes 1979). Although domestic dogs and other canids seem to be fairly resistant to plague, cats are not, and house cats can introduce plague into a family of humans by biting or scratching, or by simply breathing in proximity to a human (Kaufmann et al. 1981).

While some animals, such as domestic dogs and coyotes, are resistant to plague, prairie dogs have no resistance and succumb quickly (Gage et al. 1995; Biggins and Kostoy 2001). Once plague gets into a prairie dog colony, the mortality can be 85%–99% of the animals in the colony within a matter of days to a few weeks (Menkens and Anderson 1991). Prairie dogs have more than 20 species of fleas associated with them, and at least 5 of these have been implicated in the transmission of plague (Cully and Williams 2001). Fewer than 100 bacterial cells might be all that is necessary to kill a prairie dog, and some white-tailed prairie dogs have died when they were exposed to only two bacterial cells (Poland and Barnes 1979; Cully and Williams 2001). A single fleabite can regurgitate around 11,000–24,000 plague bacterial cells into the blood stream of a mammal (Burroughs 1947), suggesting that a single bite by an infected flea may be enough to kill a prairie dog.

Plague started to affect prairie dogs in the 1930s. The disease first appeared in Gunnison's prairie dogs in Arizona in 1932 and in New Mexico in 1938. Both Utah prairie dogs and white-tailed prairie dogs first succumbed to plague in 1936 (Eskey and Haas 1940). Plague hit black-tailed prairie dogs in Texas and Colorado in 1946 (Miles et al. 1952). Since then, plague has spread across the entire range of the Gunnison's, white-tailed, and Utah prairie dogs, and almost the entire range of black-tailed prairie dogs (Cully 1993). Some rodents, such kangaroo rats and California voles, are resistant to plague and may serve as a reservoir from which plague can expand to more susceptible populations (Holdenried and Quan 1956; Thomas et al. 1988). Because coyotes and other canids appear to be resistant to plague, one mode of transmission of plague-infected fleas from one colony of prairie dogs to another may be through fleas hitch-hiking on coyotes.

The devastating effects of plagues are so visible on prairie dogs colonies that many people erroneously believe prairie dogs are responsible for the transmission of plague. In fact, because they die so rapidly after the onset of a plague epidemic, prairie dogs have rarely been the culprit in the transmission of plague to humans (Barnes 1982). Most humans contract plague by killing, skinning, and preparing animals for consumption, while a percentage of cases are linked to domestic pets. Overall, rock squirrels and ground squirrels account for most cases of plague transmission followed next by domestic cats (Barnes 1982).

At present there appears to be no reliable mechanism to protect the prairie dog from this destructive non-native disease. Some success has been achieved by incorporating a recombinant raccoon poxvirus containing an antigen of the plague bacillus into baits that the prairie dogs could eat in the laboratory, resulting in a 56% survival among 18 prairie dogs that consumed the bait and were then infected with plague (Mencher et al. 2004). However, how to administer the vaccines to the prairie dogs under field conditions is something of a challenge. An alternative approach has been to spray burrow openings with insecticides that kill fleas. For example, permethrin, a pyrethroid insecticide, has been shown to kill fleas that transmit plague when applied at a rate of 4 grams per burrow (Beard et al. 1992). If plague is detected in time, the application of insecticides can potentially save a colony. A plague outbreak among Utah prairie dogs at Bryce Canyon National Park was halted by spraying the burrows with Pyraperm dust, a mixture that was formulated to kill fleas in the burrows of rodents (Hoogland et al. 2004).

The Biology of Success

In evolutionary terms, prairie dogs were extremely successful until humans came along. They lived in vast numbers on grasslands, coexisting with large ungulates such as bison and pronghorn antelope, and by digging burrows provided an environment for birds, other mammals, lizards, and snakes. Their color vision and acute hearing provided the basis for a sophisticated communication system that warned other prairie dogs about the approach of predators such as coyotes and hawks. Hibernation offered some of the prairie dog species a way to adapt to the harsher climate of higher elevations and northern latitudes. Living in groups provided a way to defend territories containing food resources and burrow systems. Although they bred only once a year and had relatively few young, this was a strategy that served them well in the past, ensuring the replacement of adults who grew old and died, but also ensuring that they did not breed to the point where they overpopulated their habitat. Grass on the prairies was plentiful and life was good.

We humans have changed this equation. We have destroyed the grass in large areas where prairie dogs used to live, we have classified them as pests and poisoned them, we shoot them, and we introduced a deadly disease—plague—that kills just about every prairie dog that it comes into contact with. Because our economic interests seemed to conflict with the lives of prairie dogs, we initiated a process that is leading the prairie dogs toward extinction.

Interlude:
Taxonomy and
Prairie Dog Taxonomists

There are five species of prairie dogs known to science. In this section, we would like to explore the following questions: (1) What is a species? (2) How are species named and described? (3) Who named and described the species of prairie dogs?

Probably the most widely accepted definition of a species is the biological species concept (Mayr 1969; Slobodchikoff 1976). This definition says that a species is a group of actually or potentially interbreeding populations that are reproductively isolated from other such groups. The key part of this definition is reproductive isolation. If two populations can interbreed, they are not considered to be separate species. If they cannot interbreed, they are considered to be distinct species. While this is fine from a theoretical standpoint, in practice this definition presents some difficulties. One major difficulty is in establishing reproductive isolation between two or more populations. We can often make inferences about reproductive isolation, such as when two populations might have different chromosome numbers or DNA patterns, or in the case of insects, have genitalia that are so different that reproduction between populations is simply physically impossible. Another major difficulty is that it ignores the possibility that species are evolving units, and two groups of populations might be evolving into different habitats and have different morphologies, but still may not have lost their ability to interbreed. This is the situation with several of the canid or dog species. Domestic dogs can breed with coyotes to produce dog-coyote hybrids, and can also breed with wolves to produce dog-wolf hybrids. Although wolves, coyotes, and domestic dogs live in different habitats, and have different behaviors, they have not diverged very much in terms of the similarity of their DNA and have not lost the ability to interbreed. In this case, we call wolves, dogs, and coyotes separate species primarily for our convenience, because they all look and act differently, although there is a movement among some biologists to consider domestic dogs as members of the wolf species. A third difficulty is that sometimes two groups of organisms look very similar or identical, but cannot or do not interbreed because of chromosomal differences or differences of behavior, such as activity times. Chromosomal differences among populations that have the same external morphology are found primarily in plants, but differences in activity times are found in some animals such as some insects. Still, even with these difficulties, the biological species concept offers us a criterion, most of the time, for deciding when two or more populations all belong to the same species.

Historically, species were identified on the basis of morphological characteristics. If two specimens were different in their physical traits, taxonomists (taxonomy is the science of describing and classifying organisms) generally considered them to be separate species. In the 1800s, when there were a large number of naturalists exploring the world and sending specimens of animals and plants to people working on classifying the natural world, taxonomists fell into two groups: the lumpers and the splitters. Lumpers were people who believed that each species potentially exhibited a wide range of morphological variation, and that small deviations in physical traits between two specimens did not necessarily indicate that these specimens represented two separate species. Splitters, on the other hand, thought that morphological differences between specimens often reflected membership in separate species.

Whether described and named by lumpers or splitters, species are then placed into genera (singular, genus), which is a higher category that brings together groups of similarly related species. By convention, both genus and species names are italicized, and the genus name always starts with a capital letter while the species name always starts with a small letter, even if the species is named after a person or a place.

Both lumpers and splitters went about the process of identifying new species in similar ways. They would find one or more specimens of something that they thought was new to science and write a description. This description was an attempt to provide as thorough a word picture of the specimen as possible, describing the color, size, measurements of anatomical parts, and the location of where the specimen was found. This description was then published in a book, a monograph, or a scientific article, and the species was then considered named and described by the author of the publication. In more recent times, one specimen would be designated as the holotype, or type specimen, which would be deposited in a museum where other taxonomists could look at it as a reference point for what the author of that species name considered to be the species. Eventually the whole process was formalized into a series of rules, called the International Code of Zoological Nomenclature (just for animals; plants have their own code), that spells out the procedures and steps for naming and describing a species that is new to science (www.iczn.org). These are the rules that taxonomists have to follow today in order to for a new species to be considered a valid name and to be accepted by scientists worldwide.

Once a species has been named and described, the name of the author who wrote the description is often appended to the species name. For example, the black-tailed prairie dog is placed in the genus *Cynomys* (which means dog-mouse) and the species name is *ludovicianus*. Taxonomists added the name Ord, and the date 1815, so that the full genus and species name of the black-tailed prairie dog is *Cynomys ludovicianus* Ord 1815. This means that a taxonomist with the last name of Ord described this species in 1815. The name of the author of the species description and the date are often useful because taxonomists can move a species from one genus into another as new information about taxonomic and evolutionary relationships becomes available. The name of the author and the date of the description move along with the species, and help prevent confusion if there are other species with the same name in other taxonomic groups; other similarly named species probably were described by a different author at a different date.

All of the five species of prairie dogs are members of the genus *Cynomys*, which was named and described as a genus in 1817 by Constantine Samuel Rafinesque. Rafinesque also described a species of prairie dog in this new genus, *Cynomys socialis*, which was subsequently found to be the same as the species *ludovicianus* described by George Ord two years earlier in 1815 (taxonomists refer to such names as synonyms and the process of identifying different names that refer to the same organism as synonymy). However, Ord described the black-tail prairie dog species *ludovicianus* as a member of another genus, *Arctomys*, which was the genus in which marmots were included until the marmot genus name changed to *Marmota*. Because Ord described the species *ludovicianus* in 1815, two years before Rafinesque described the same animal as *socialis*, Ord's species description is considered to be the valid one (this is called priority in taxonomy), even though he described it in a different genus. The species *ludovicianus* was moved to the genus *Cynomys* in 1858 by Spencer Fullerton Baird. Baird was also the first to name and describe the Gunnison's prairie dog, *Cynomys gunnisoni*. He published his description of this species in 1855 but thought that it belonged in a different genus of ground squirrels, the genus *Spermophilus*. In 1858 he decided that *gunnisoni* was really a prairie dog, and changed the genus designation to *Cynomys*. The white-tailed prairie dog species *leucurus* was named and described by C. Hart Merriam in 1890, and the Mexican prairie dog species *mexicanus* was also named and described by Merriam in 1892. The Utah prairie dog species, *parvidens*, was named and described by Joseph Asaph Allen in 1905. By the time Merriam and Allen were describing species, the prairie dog genus *Cynomys* was well established, and all three species described by these two people were placed in this genus.

Of the five naturalists-taxonomists mentioned above, Rafinesque (1783–1840), who gave us the genus name of prairie dogs, *Cynomys*, was a splitter. During his lifetime he described some 2,700 genera and some 6,700 genus and species combinations (Merrill 1949). He was born in Turkey and grew up in Marseilles, France (Reveal 2003; Kimberling 2006). He had little formal schooling, but he had read a thousand books and taught himself Latin by the age of 12. His passion was plants. As a child he built a herbarium, and he wanted to spend his life studying plants. In an age when people generally had some profession or occupation other than as a naturalist, Rafinesque came to America at the age of 19 and apprenticed at a mercantile house in Philadelphia. In 1804 he asked Thomas Jefferson to be appointed as a naturalist for an expedition to the American West, but Rafinesque had to leave for Sicily on a business venture before the appointment came through. He stayed in Sicily for 10 years, where he collected plants, wrote manuscripts, and accumulated books. He published articles in scientific journals where he described a novel system of classification of plants and animals as well as naming new species of plants. On his return to America, all of his collections, books, and manuscripts were lost at sea as the ship that was carrying them sank off the coast of Connecticut. With little money and few possessions Rafinsque went to New York where he found a wealthy patron, Dr. Samuel Latham, who took him into his home and found him a job at the

Lyceum of Natural History. Rafinesque traveled widely through the Allegheny Mountains, collecting more than 250 new species, and published numerous articles in scientific journals. Partly because he was a splitter and provided few details of the new species that he named, other botanists ignored much of his work. By 1819 Rafinesque was appointed professor of botany at Transylvania University in Kentucky where he also taught French and Italian and was the university librarian. From 1819 until 1825 he published several books with descriptions of new species, and described 66 new genera of plants in North America. However, other botanists disagreed with the fine-scaled distinctions that Rafinesque used in his descriptions, and once again his work was largely ignored. Unfortunately, he did not feel that he had to be present at his classes on a consistent basis, so he was released from his job in 1826. Taking all of his collections, manuscripts, and books, he moved to Philadelphia where he lived out the remainder of his life continuing to write articles and describe new species and genera, which were largely ignored by other scientists and naturalists and caused him considerable frustration and bitterness. He died penniless from stomach cancer, and because no one thought that his work was very valuable, his collections and unfinished manuscripts were largely lost or destroyed through the ravages of time.

The person who described the black-tailed prairie dog, George Ord (1781–1866) was much more successful as a scientist and naturalist than Rafinesque (Peck 2000). Ord grew up in Philadelphia, and at the age of 19 he joined his father's rope-making business. He continued in this business for 29 years taking over the family firm when his father died. In 1829 he retired from making ropes so that he could pursue his scientific studies full-time. During the time he was working in the family business, Ord spent part of his time on scientific pursuits. In 1805 Ord became friends with the naturalist Alexander Wilson, who was writing a series of books called *Ornithology: The Natural History of the Birds of the United States*. When Wilson died in 1813 leaving the work unfinished, Ord finished the eighth volume of *Ornithology*, and then wrote the ninth and last volume in 1814. Also interested in mammals, he intended to publish a series of volumes about North American mammals, but had to give up the project because of the expense involved. Throughout this time he authored numerous articles in scientific journals, and was generally considered to be an important figure in the development of an understanding of the natural history of America. During his career he was an officer of the

American Philosophical Society, a fellow of the Linnean Society of London, and president of the Academy of Natural Sciences of Philadelphia. In general, he was not well liked. He had a temper, and could often use his facility with words to cause other people distress. He detested John James Audubon, and spent a considerable amount of time telling everyone that Audubon's work was insignificant and inaccurate, calling him a charlatan and an impostor so that many scientists were initially reluctant to accept Audubon's work on the strength of such poor recommendations. As an irony of history, Audubon's name is widely known today, while Ord is rarely remembered. Toward the end of his life, he became a recluse spending his time with his books and shunning people.

Another successful scientist and naturalist was Spencer Fullerton Baird (1823–1887), who named and described the Gunnison's prairie dog *gunnisoni* in 1855. Baird grew up in Pennsylvania and attended Dickinson College from 1836 to 1840, during which time he developed a passion for natural history, collecting birds and other animal specimens (Goode 1996; Jackson 2000). After college he continued to collect specimens and continued to study natural history, taking drawing lessons from John James Audubon, although he started studying medicine so as to have a profession. When he was 18 he went on a walking trip through the mountains of Pennsylvania to collect and observe birds, walking 400 miles in 21 days. At the age of 20 he published a paper describing two new species of flycatchers. When he was 22 he was appointed professor of natural history at Dickinson College, and at age 23 became chair of natural history and chemistry at that college. This was the start of a long career in what today we would call science administration. Although Baird was granted an honorary MD in 1848, he never worked in the medical field. In 1850 Baird was appointed assistant secretary and curator at the Smithsonian Institution. He brought his collections of specimens with him, and these collections served as the nucleus for the natural history collections at the U.S. National Museum. At the Smithsonian, Baird was influential in sending collectors to various part of the world to increase the holdings of the museum collections, and began exchanging specimens with other museums, building extensive collections particularly of birds and fish, two major interests in Baird's life. Although he had extensive administrative duties, he also published a series of books on a variety of animals, including *Catalogue of North American Serpents* (1853), *Birds of North America* (1858), and

Mammals of North America (1859). In 1878 he was appointed as the second secretary of the Smithsonian Institution (the secretary is the chief executive officer), and served in that capacity until 1887, when he retired. During his life, he received numerous awards, both national and international, and had many species named after him, ranging from invertebrates such as starfish, mollusks, and butterflies, to snakes, birds, and mammals, and even had a post office named after him in California in 1877.

Just as Baird was a major figure in American natural history, so was C. Hart Merriam (1855–1942), who named and described the white-tailed prairie dog species *leucurus* in 1890 and the Mexican prairie dog species *mexicanus* in 1892. Merriam was born in a rural mansion in the Adirondack Mountains of New York (Osgood 1943; Sterling 2000b). As he was growing up, he had ample opportunity to observe the plants and animals around him. By the time he was 15, he had started a collection of birds and at the suggestion of Spencer Baird, whom Merriam met when his father took him to Washington, DC, was learning how to prepare scientific specimens. By the time that he was 17, he went along on a government-sponsored expedition to Yellowstone, collecting some 313 bird specimens and 67 nests and eggs. As his first scientific article, he wrote a report of his collecting efforts, providing an annotated list of the specimens that he collected on that trip. In Merriam's recollections, this trip had a strong effect on his subsequent career. On the trip, he met a number of other young naturalists who became lifelong friends, and also developed a lifelong interest in the American West. Although he was intensely interested in natural history, he was expected by his family to have a more lucrative profession, so he spent four years at Yale University, studying medicine, and then went on to Columbia University, where he received an MD when he was 24 years old. During this time he kept up his interest in birds, publishing several articles on the birds of Connecticut and New York. From 1879 to 1885 he practiced medicine in Locust Grove, New York, the town in which he was born, continuing his interest in birds and developing an interest in mammals as well. In 1884 he published the book *Mammals of the Adirondacks*, which provided a series of life histories of mammals compiled from his own observations as well as written accounts by others. His interest in mammals continued to grow. He started accumulating a large collection of mammal specimens, something unusual at the time because most naturalists preferred to collect birds, and bird collections were fairly numerous, but mammals were rarely collected. At that time, relatively little was known about mammal species both in America and in the rest of the world. Mammal specimens were often poorly prepared, had little information associated with them, and mostly represented diurnal species such as rabbits that were easy to find as collectors chased the rare and difficult-to-find bird species. By 1884, Merriam had accumulated a collection of mammals totaling some 7,000 specimens, more than any public collection at the time.

Merriam's career veered away from medicine in 1885 when he was offered the position of Ornithologist in the Division of Entomology within what now would be the U.S. Department of Agriculture. This was probably at the urging of Spencer Baird, who had been following Merriam's work and thought highly of him. Merriam accepted the position, and a year later parlayed this position into an independent Division of Ornithology, which morphed into the Division of Ornithology and Mammalogy, and then later in the 1890s into Bureau of Biological Survey, a precursor to today's Fish and Wildlife Service. Merriam stayed as head of these organizations for 25 years, battling Congress for appropriations and training and sending out investigators to study the birds and mammals of America. However, Merriam was not one to remain chained to a desk. Many of his summers were spent traveling through the American West. A seminal trip to the West was an expedition to the San Francisco Peaks near Flagstaff, Arizona, where he developed the life-zone concept that essentially states that both plants and animals are distributed in distinct life zones characterized by climatic or altitudinal factors. From what he saw on the San Francisco Peaks, he extended the concept to all of North America, drawing life zones for different latitudes and different habitats. He became interested in studying not only the classification of mammals, but also their distribution, ecology, and anatomy. He cultivated young and ambitious naturalists, spending time with them developing rigorous methods, reviewing and revising manuscripts, and discussing new ideas. Merriam was not particularly interested in general literature, art, music, or current affairs, and was not very good at dealing with congressmen who were responsible for appropriations for his bureau, depending on friends to help him out when he ran into political difficulties in getting sufficient money. In 1910 Merriam's friends persuaded Mrs. E. H. Harriman to establish a Harriman Trust that specified that Merriam was to be supported in his research endeavors for the rest of his life. Merriam left his government position and spent the rest of his life studying the anthropology of California Native Americans, whose traditions he wanted to document before they vanished.

The last of the prairie dogs, the Utah species *parvidens,* was named and described by Joel Asaph Allen (1838–1921) in 1905. Allen was born in Massachusetts into a family of farmers (Chapman 1930; Sterling 2000a). He developed an interest in nature at an early age and began collecting specimens of plants and animals even though at first he had no books on preparation techniques and had no one to turn to for advice. In his mid-teens, a teacher at his district school shared his enthusiasm for nature and encouraged and helped him with his studies of the natural world. Then, from 1858 to 1862, Allen attended the Wilbraham Academy in the winters, when he wasn't needed on the family farm, and studied natural history with another teacher, collecting some 100 species of birds, mammals, fish, and insects during the four years that he was there. In 1862, Allen went to the Lawrence Scientific School at Harvard University to study with Louis Agassiz, where he stayed until 1865. Allen was then invited by Agassiz to join him on a year-long expedition to Brazil. Unfortunately, chronic health problems had been plaguing Allen most of his life, and after the expedition, Allen left Cambridge to go back to the family farm to recover from a bout of chronic indigestion. He returned again to the Museum of Comparative Zoology at Harvard in 1867 as curator of bird and mammal collections. In 1871 Allen set out with a couple of assistants on a nine-month expedition to the Great Plains and Rocky Mountains. The expedition brought back 200 mammal skins, 60 skeletons, 240 skulls of most large mammals, 1,500 bird skins, more than 100 birds in alcohol, and a variety of insects, mollusks, and fossil fish. The next year, Allen set out on another 550 mile-long expedition that went between Bismarck, South Dakota, and Yellowstone, accompanied by 1,400 troops commanded by General Custer. The expedition was attacked twice by hostile parties, and the naturalists were forbidden to use guns or to stray away from the troops in case they would precipitate another attack. After this, Allen's health was greatly impaired, so he stopped going on any further expeditions, spending his time instead on research in museums. He accepted a position as curator of birds and mammals at the American Museum of Natural History in 1885, where he spent the next 32 years, publishing 37 papers on birds and 165 papers on mammals. In addition to describing new species, he was also interested in geographic variation among mammals and birds. He first observed what later became known as

Allen's Rule, that animals in cold climates tended to have smaller extremities as a way of conserving heat. On a personal level, Allen has been described as being modest, unselfish, considerate of others, but at the same time rigorous in scientific thinking, critical of careless work or conclusions based on insufficient data. He was a long-time editor of the journals the *Nuttall Bulletin* and the *Auk*, which succeeded the *Nuttall Bulletin*. He won many awards and honors, which his modesty always led him to describe as surprises.

The common characteristics that the above five naturalists-taxonomists had was a great love of nature and a great sense of curiosity about the natural world. They lived in a time when much was unknown about the plants and animals of the world, and the way that knowledge was accumulated was through collecting specimens. Each person spent a considerable amount of time either accumulating large collections, or studying collections in museums, or both of these forms of activity. Each loved what they did. Today there is much less emphasis on collecting specimens, and much more emphasis on studying the behavior, ecology, and functional anatomy of animals. Taxonomy has been subsumed in the larger field called systematics, which attempts to look at broad-scale relationships between species and higher-order categories such as genera, orders, families, and phyla. The tools of the systematist now are DNA analysis, statistics, and sophisticated microscopy, rather than the hand-lens and ruler of the early-day taxonomists. On the basis of similarities and differences in DNA structure, we now have much better bases for including species into a genus and for understanding the evolutionary relationships between species that look similar morphologically. However, a huge contribution of the five naturalists discussed earlier, as well as many other taxonomists then and now, is that they have provided us with names and descriptions of species, and these names serve to organize behavioral, ecological, and anatomical information about the animals in question. Without the names, we would have nothing to talk about.

3

The Social Behavior of Prairie Dogs

Prairie dogs have a complex social system, rivaling that of some primates and in some respects resembling the behavior of humans. Perhaps that is why prairie dog exhibits in zoos are very popular, with people sometimes spending hours watching the behavior of the animals. They live in social groups that have been called coteries (King 1955; Hoogland 1995), occupying territories with well-defined boundaries (Slobodchikoff 1984). On a relative scale of sociality, black-tailed and Mexican prairie dogs are considered to be more social while Gunnison's and Utah prairie dogs are somewhat less so, and white-tailed prairie dogs are least social. This social scale is based in part on the interactions of animals within their social groups. Black-tailed and Mexican prairie dogs have a considerable amount of interaction with frequent amicable behaviors toward other members of the social group, while the other species have fewer such interactions. These interactions include the greet-kiss, where two prairie dogs come toward one another, open their mouths, and press their tongues together. After doing this, the two prairie dogs go back to whatever they were doing previously if they are members of the same social group. If, however, they do not belong to the same group, then one prairie dog will chase the other one out of its territory after the greet-kiss. Other interactions include allogrooming, or mutual grooming, which is done by the black-tailed and Mexican species, but not the Gunnison's, Utah, or the white-tailed prairie dogs.

In this chapter, we give an overview of the social systems of the five species, showing points of commonality and difference. We discuss how in at least one species, the Gunnison's, the social system seems to relate to the distribution of food resources, and the social system is not the same as the mating system. We also discuss some of the specific behaviors that occur within the

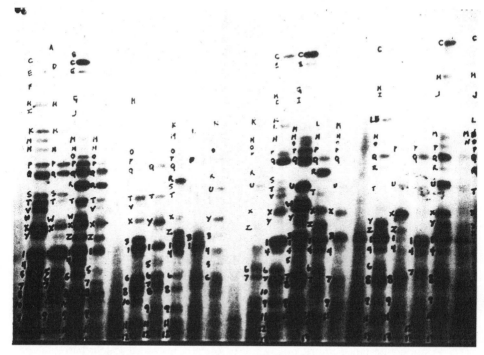

Figure 3.1. Example of a DNA gel used for determining paternity and relatedness in
        Gunnison's prairie dogs. Each column of black bands represents a different
        animal. Each band represents a different DNA segment of different lengths,
        allowing the comparison of genetic similarity between different individuals.

social systems, such as territorial defense, infanticide, greet-kisses, mutual
grooming, and communal nursing.

    In our discussion of the social systems of the different species, we use the
terms socially monogamous and socially polygynous, where monogamous
means one adult male mates with one adult female, and polygynous means
that one adult male mates with two or more females. Animal behaviorists
have often assumed that the mating system is the same as the social system—
for example, when we say that a human society is monogamous, we assume
that the social families within that society live in monogamous groups of
a single male and a single female, and also mate monogamously with each
other. However, at least for the Gunnison's prairie dog, the mating system
is not the same as the social system. Genetic studies involving DNA finger-
prints of Gunnison's prairie dogs do not support that idea, showing that

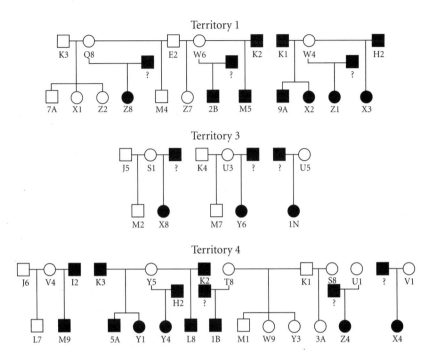

Figure 3.2. Pedigree for Gunnison's prairie dogs at colony Antelope Hill for three different territories. Letters and numbers identify individual animals. Black shading denotes extraterritorial males and their offspring. (From Travis et al. 1996.)

Gunnison's females have multiple paternity—like cats, the females can have the pups sired by multiple males within a litter. When she is sexually receptive (which happens for only a few hours on a single day) a Gunnison's female apparently mates with multiple males, some of whom live on her territory, but most who do not. The DNA analysis found that on average 67% of the pups were sired by extraterritorial males, i.e., males that live on territories other than the one occupied by the female (Figures 3.1–3.3). On average, a resident male sired only one of the pups in a resident female's litter of two to five (Travis et al. 1996). This means that not all the pups living on a territory are genetically related to the resident male—in fact, the resident male may have more pups related to him living on other territories. Another study that assessed the frequency of multiple paternity in litters, but did not take into account whether the males came from other territories, found that 77% of the Gunnison's prairie dog litters showed multiple paternity, and 71% of the

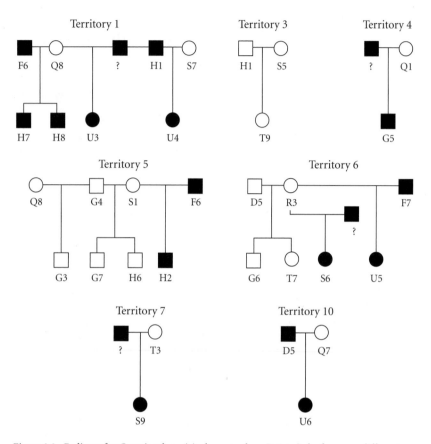

Figure 3.3. Pedigree for Gunnison's prairie dogs at colony Potato Lake for seven different
territories. Letters and numbers identify individual animals. Black shading
denotes extraterritorial males and their offspring. (From Travis et al. 1996.)

litters sampled one year and 90% of the litters sampled the following year
were sired by two or more males (Haynie et al. 2003). An earlier genetic study
of black-tailed prairie dogs suggested that a resident male sired many of the
pups within a territory (Foltz and Hoogland 1981)—however, this analysis
was done before the precision of DNA technology was available.

Mating, Relatedness, and Territory

If females leave their territory to copulate with other males, we can ask, what
is the advantage to the females? A study of Gunnison's prairie dogs at Petri-

fied Forest National Park, Arizona, found that there are two distinct advantages for a female to mate with multiple partners (Hoogland, 1998a). One is the probability of conception and giving birth. Females who mated with only a single male had a probability of 92% of giving birth to a litter, while females who mated with three or more males had a 100% probability of giving birth. The other advantage is that females who copulate with five males have an average litter size of approximately 4.7 pups, while females who copulate with a single male have an average litter size of 3.5 pups.

Copulation in prairie dogs takes place below ground and is difficult to observe. However, there are some behavioral indications that copulation is taking place between a male and a female. In both black-tailed and Gunnison's prairie dogs, one of the first behaviors that can be noticed is that a male and a female disappear together into a burrow for varying time intervals ranging between 15 seconds and one day (Hoogland 1995, 1998b). Males have a very limited time interval in which to mate with a female, because female prairie dogs are in estrus and receptive to mating for only a single day (Hoogland 1995). Although a male and a female can enter a burrow together at any time of day, most such entrances occur in the late morning or the afternoon. Prior to copulation, a male sometimes spends some time sniffing the anal area of the female, and after copulation, the male can attempt to guard the female from having access to other males. The male also might take a few mouthfuls of nesting material (usually dried grass) into the burrow. After copulation, both males and females can engage in self-licking of their genitals, and also can engage in dust bathing. Both black-tailed and Gunnison's males can give a "mating call," which superficially resembles an alarm call (Hoogland 1995). In his studies of the mating behavior of prairie dogs, Hoogland (1995, 1998a) found that other than the behavior of disappearing into a burrow, none of the other behaviors occurred in very high frequencies (with a range of 9%–47% of individuals observed exhibiting a particular behavior), perhaps because some of these pre- and post-copulatory behaviors take place mostly below ground. The most frequent behavior that Hoogland found above ground was the "mating call," which was given by 57% of the black-tailed males and 54% of the Gunnison's males at the presumed conclusion of copulation.

Although it has been assumed that black-tailed prairie dog social groups (coteries) are family units, the genetic data to support this are still lacking. Using behavioral indications, one study concluded that only 16% of females copulate with a male from outside their territory (Hoogland 1995). These behavioral indications involved the indications listed above, i.e., noting that a particular male was attentive to a female, engaged in nest-building activities such as carrying dried grass into a burrow, gave a mating call, and licked his genitals after the supposed copulation. However, because copulation in prairie dogs occurs primarily below ground, where it is impossible to see what actually happens, behavioral data can serve as the basis for inferences about who mated with whom, but genetic data about paternity and maternity provides the actual evidence. Some genetic evidence suggests that the animals within a coterie are highly related (Chesser 1983), but the problem with generalizing this evidence to conclude that coteries are family groups is that populations of prairie dogs may have high levels of relatedness within the entire colony (Travis et al. 1996). For example, the level of relatedness in Gunnison's prairie dogs of any two animals in the colony, living in separate territories, is $r = 0.55$ (where r is the coefficient of similarity, representing the average proportion of DNA fragments shared among individual prairie dogs within the colony; when $r = 0$ there is complete dissimilarity, and when $r = 1$ there is complete similarity). All the prairie dogs on the Gunnison's colony are highly related, but the social groups do not represent family groups in terms of paternity of the pups (Travis et al. 1996).

If all the animals in a colony are highly related, the big question then is: Why live in a social system in the first place? Why not live in a hive-like community, perhaps like bumblebees or like social wasps? The answer seems to lie in the distribution of food resources, and the social system of Gunnison's prairie dogs gives us some clues to the solution to the above question. Gunnison's prairie dogs have a flexible social system, i.e., they can span the entire range from socially monogamous to socially polygynous to multi-male/multi-female assemblages (Travis and Slobodchikoff 1993). All of these social systems can be seen in different territories within a single colony (Rayor 1988; Travis et al. 1995, 1996, 1997), although more typically an entire colony has one predominant type of social system (Travis and Slobodchikoff 1993) (Figure 3.4). A single social assemblage occupies a single territory, which is cooperatively defended by all the members of that social group (Slobodchikoff 1984; Rayor 1988). Some colonies have most of the territories occupied by socially monogamous groups, while other colonies have most of the territories occupied by socially polygynous or multi-male/

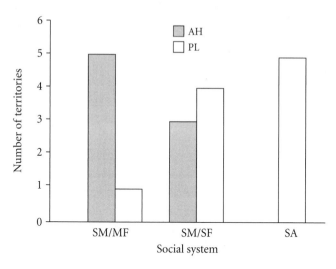

Figure 3.4. Variation in social systems within territories in two colonies of Gunnison's prairie dogs, Antelope Hill and Potato Lake, near Flagstaff, Arizona. SM is single male, SF is single female, MF is multiple females, and SA is single adult. (From Travis and Slobodchikoff 1993.)

multi-female groups, where a multi-male/multi-female group is one that has several males and several females occupying a single territory (Travis and Slobodchikoff 1993).

Gunnison's prairie dogs have a mean territory size that is similar to that of the black-tails. A study of Gunnison's in an arid environment (Petrified Forest National Park, Arizona) found a mean territory size of 0.67 hectares (1.7 acres) (Hoogland 1999), while a study in a more mesic environment near Flagstaff, Arizona, found a mean territory size of 0.30 hectares (0.75 acres) (Travis and Slobodchikoff 1993). The mean size of a territory of black-tails in Wind Cave, South Dakota, was 0.31 hectares (0.78 acres) (Hoogland 1995).

Territorial defense is often cooperative. In Gunnison's prairie dogs, all of the animals within a territory, both the males and the females, can participate in chasing out intruders (Arjo 1992). In black-tailed prairie dogs, territorial defense starts when an intruder strays across a boundary into another territory (King 1955). Upon seeing the invading animal, a resident rushes to confront the intruder. If the invader does not run away, both animals stop and face each other, at a distance of 1–8 meters apart. One of the animals, either the intruder or the resident, fluffs out and elevates its tail, turns and exposes its anal glands. The other animal approaches slowly and sniffs the glands. Then the other animal turns and allows its anal glands to be sniffed. This sniffing behavior can alternate from one animal to the other, until one of the animals tries to bite, or the intruder runs away. The resident animal can produce a series of vocalizations that attract other individuals from its territory to help with the defense. These territorial disputes can occur between both males and females. At his study colony of black-tailed prairie dogs in Wind Cave National Park, South Dakota, King (1955) noted that in one day of observation, he saw 17 disputes between members of different territories. Of these, three were between two males, five were between two females, and nine were between a male and a female.

Within a territory, the resident animals can engage in both amicable and aggressive behaviors. Amicable behaviors involve a greet-kiss (see page 56) (Figure 3.5), or feeding side-by-side. Aggressive behaviors involve facing each other and fluffing out the tails, or displacements by one animal of another from where it was feeding, or chases or fights. A study of intra- and inter-territory aggression in Gunnison's prairie dogs found that aggressive behavior tended to occur as frequently within a territory between the members of that territorial group, as occurred between territories, and neither type of behavior occurred very frequently (Arjo 1992). In 98 hours of observations of a colony over a 22-week period from mid-August through September, Arjo (1992) recorded 10 aggressive interactions within the social groups, and 6 aggressive interactions between social groups, a difference that was not statistically significant. Among the Gunnison's females that were studied within their territories, there did not appear to be any evidence of dominance hierarchies. Of the six female groups that were examined, four were characterized by having entirely amicable interactions. Among the remaining two groups, most of the females had amicable interactions with one another within their groups, but in one group two out of the six females were targets of aggressive behavior by the other females, while in the other group one female out of the three was the initiator of aggressive behavior directed against the others (Arjo 1992). King (1955) noted that most of the interactions within the territorial groups of the black-tailed prairie dogs that he studied were amicable.

Although the territorial boundaries remain fairly stable from year-to-year (King 1955; Hoogland 1995), there can be movement of individuals from one territory to another within a colony. Among the black-tails, prairie dog males

Figure 3.5. Greet-kiss by two black-tailed prairie dogs. (Photo by C. N. Slobodchikoff.)

tend to leave their natal territories as yearlings, before they become sexually mature. Hoogland (1995) reports that out of 312 juvenile males, not a single one was left on its natal territory by the Spring of its fourth year. In contrast, black-tailed females tend to be philopatric (remain on the territory where they were born). Out of 239 juvenile females, Hoogland (1995) found that only 11 moved to another territory, and all of these moved by the time that they were two years old. Hoogland (1995) also found that adult males tend to move from one territorial group to another after about two to three years.

Gunnison's prairie dogs apparently have a more complex pattern of move-ment to other territories within their colony, in which the numbers of adults and yearlings that move to other territories varies from year-to-year. In a study of two colonies of Gunnison's prairie dogs in the vicinity of Flagstaff, Arizona, Robinson (1989) found that males and females, adults and year-lings, move to other territories. In one colony, 75% of the adult males, 20% of the adult females, 40% of the yearling males, and 8% of the yearling females moved to another territory during a single field season. In the other colony, in one year 82% of the adult males, 40% of the adult females, 90% of the year-ling males, and 5% of the yearling females moved to another territory, while in the second year of the study, 70% of the adult males, 20% of the adult females, 38% of the yearling males, and 42% of the yearling females moved. Most of these movements were in the spring. Although both adult males and yearling males moved more frequently than females, a significant number of females of both age classes moved as well. Interestingly enough, Robinson (1989) found that there was no significant difference in subsequent reproductive success among the females that moved to new territories, compared with the reproductive success of females that were philopatric and stayed on their natal territories. In a study of a different colony of Gunnison's prairie dogs, Arjo (1992) found that none of the nine adult males stayed on the territory they had occupied in the previous year, while 13 of 16 females occupied the same territory that they had the year before.

One suggestion for why the males move from their territories is that this is a mechanism of reducing inbreeding. Hoogland (1995) has suggested that males may leave their territory when their daughters, who tend to be philopatric and stay in their natal territory, become sexually mature.

Social Systems

Although black-tailed prairie dogs have been described as socially polygynous (King 1955; Hoogland 1995), the evidence suggests that they too have a flexible social system. Even though most of the social groups that have been studied fit the description of polygyny (one breeding male and two or more breeding females), not all social groups of black-tails have this kind of social structure. At Wind Cave National Park in South Dakota, some 27% of the social groups contained multiple males (Hoogland 1995). Also, studies of black-tails at Wind Cave describe some social groups as having one adult male and one adult female, or a single animal of either sex occupying a territory (King 1955). In a study of black-tails near Fort Collins, Colorado, four of the eight coteries that were studied had two, three, or four males in a coterie, with either one or zero females. The remaining four coteries had one male and one female. These social groups occupied territories that were stable in their boundaries from one year to the next (Tileston and Lechleitner 1966).

In Gunnison's prairie dogs, the structure of the social group is related to the food resources on a territory (Slobodchikoff 1984; Travis and Slobodchikoff 1993; Verdolin 2007) (Figure 3.6). When food plants are distributed fairly uniformly and are abundant within a territory, the social group occupying

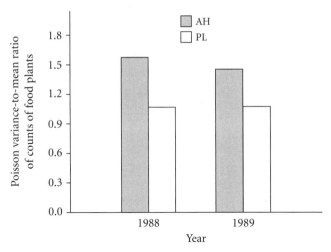

Figure 3.6. Variation in the distribution of Gunnison's prairie dog food plants at two
        colonies, Antelope Hill and Potato Lake, near Flagstaff, Arizona. A Poisson
        variance-to-mean ratio near 1.0 indicates a uniform distribution of food plants,
        while a ratio higher than 1.0 indicates a patchy distribution of food plants.
        (From Travis and Slobodchikoff 1993.)

the territory tends to be monogamous, i.e., one adult male and one adult
female. However, when food plants are patchily distributed, the social group
occupying the territory tends to be socially polygynous, i.e., one male and
several females, or multi-male/multi-female. A model for how this kind of
flexibility of social structure can arise suggests that when food plants have a
uniform distribution, two animals can easily defend all the food resources that
they need. However, when food plants are patchily distributed, two animals
may not be able to defend all the plants that they might need to eat (here the
choice of plants might relate to nutritional as well as energetic factors). So,
additional animals might get incorporated into the social group, up to the
point where there are enough animals to defend the food resources within the
territory and satisfy the nutritional demands of each animal occupying the
territory. If the food plants were patchily distributed throughout, the entire
colony would have socially polygynous or multi-male/multi-female territorial groups (Slobodchikoff 1984) (Figure 3.7). Although the ecology of black-
tailed prairie dogs has not been studied in this respect, black-tailed social
groups might respond to resources in the same way, accounting for the variation that has been observed in social systems.

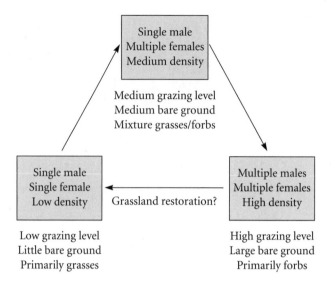

Figure 3.7. Hypothetical scheme for the relationship between social system, density of
          prairie dogs, and grazing levels by large ungulates. A possibility that remains
          to be tested is the restoration of grasslands to lower densities of prairie dogs by
          controlling grazing by large herbivores.

Among the Gunnison's prairie dogs, the variation in social system is often
associated with habitat, which in turn is associated with food plant patchiness. Habitats of grasslands that have not been overgrazed represent fairly
uniform distributions of food plants, while habitats of grasslands that have
been severely overgrazed—either by the prairie dogs themselves or by large
herbivores such as cattle—represent patchy distributions of food plants,
with considerable patches of bare ground between patches of food plants. As
the plant patchiness increases and the proportion of bare ground increases,
the density of the prairie dogs also tends to increase (Travis et al. 1995). In
severely overgrazed colonies, there is often nothing but bare ground left, yet
the densities of prairie dogs are high, and the social systems are predominantly multi-male/multi-female.

So, the answer to the question of why prairie dogs live in social groups is
that they might do this to exploit their resources. Each social group defends
food resources that are crucial to the survival of the individuals living in the
territory. In addition to the food resources above ground, each social group
defends a burrow system that is also crucial to the survival of the individual

group members. Collectively these resources promote the individual fitness of each animal living in the territory, and each animal has a stake in cooperatively defending those resources.

A question often arises, how do the prairie dogs manage to live in these colonies with little but bare ground available? We speculate the answer is related to distributions and availability of seeds in the colony. Gunnison's prairie dogs include a considerable proportion of seeds in their diet (Shalaway and Slobodchikoff 1988). Seeds are packets of high-energy foods—a sunflower seed might have twice the caloric value of a grass stem. Gunnison's prairie dogs eat seeds early in the spring, when there is no green vegetation present, then switch to grasses and forbs when the rains come and the plants start to grow, and then they switch back to seeds in the fall, when the green plants start to die back. In habitats where there is almost continuous grass cover, the prairie dogs have limited access to the buried seeds, because the grass stems and roots keep the animals from digging up the seeds. In habitats where there is a considerable amount of bare ground, however, the prairie dogs can easily have access to the seeds stored away in the soil, which serve as high-energy packets of food for their daily activities. The seeds are distributed very patchily (Reichman 1984; Price and Reichman 1987), and this in turn might influence the social groups to be multi-male/multi-female because a larger group might be necessary to defend the patchily-distributed resources. Black-tailed prairie dogs also eat seeds (Fagerstone et al 1981), and it is possible that seeds can support populations of these prairie dogs that live in areas where there is no apparent food for them.

Although we need more data on this point, there might be a relationship between the social system, patchiness of food resources, and seeds in the habitat. When grass cover is continuous and fairly uniform, there are relatively few open patches where the prairie dogs can get access to seeds, so they feed on the grasses and forbs, which contain less energy than the seeds. The social systems at this point consist of monogamous pairs or single animals. As large herbivores start to graze more heavily on the grasses within a colony, their grazing opens up more patches of bare ground, allowing the prairie dogs to have more access to seeds. The higher energy food allows the density of prairie dogs to go up, and the patchiness of the food resources shifts the social systems to polygynous or multi-male/multi-female (Travis and Slobodchikoff 1993; Travis et al. 1995) (Figure 3.7).

Unlike the Gunnison's and black-tailed species, the social structure of the Mexican, Utah, and white-tailed prairie dogs is not extensively documented but also appears to be socially flexible. The Mexican prairie dogs are assumed to have a social system much like the black-tailed, while the Utah and white-tailed prairie dogs have been described as being somewhat less social. In a study of white-tailed prairie dogs near Walden, Colorado, the animals were observed over two years, 1959 and 1960 (Tileston and Lechleitner 1966). Of the six territories that were observed, one territory was socially monogamous and one was socially polygynous in both years. Of the remaining four territories, one had only a single female and no male in both years, one went from socially monogamous to multi-male/multi-female (two males, three females), one territory went from a single female and no male in one year to no animals at all in the following year, and another territory was set up in the second year by one male and two females. As with the other species, these results suggest a considerable amount of variation in the social system.

The Greet-Kiss

Within a territory, prairie dogs engage in an amicable behavior known as the greet-kiss or greeting-kiss (King 1955; Steiner 1975). Why they do it is unknown. Two prairie dogs will come toward each other, open up their mouths, and press their tongues together for a brief period of time. This kind of behavior is seen in other ground squirrels as well. Greet-kisses can occur between two males, between two females, between a male and a female, between adults and juveniles, or between two juveniles (King 1955; Fitzgerald and Lechleitner 1974).

The functions of a greet-kiss are not fully known, but several hypotheses have been suggested. One hypothesis is the Food Information Hypothesis (Steiner 1975), where it is suggested that information about the kind of food a prairie dog has eaten, its smell and taste, can be passed on from one animal to another during an exploration of each other's oral cavity. Another is the Individual Recognition Hypothesis (King 1955; Tileson and Lechleitner 1966; Steiner 1974; Owings and Hennessy 1984; Ferron 1985), which suggests that greet-kisses might function in recognizing individual animals. Although prairie dogs have excellent vision and might perhaps be able to recognize visually other individuals aboveground, identifying individuals belowground, where everything is dark, by their smell or taste might have distinct advantages. A third hypothesis is Dominance Maintenance (Anthony 1955; Steiner 1975), which suggests that prairie dogs might use greet-kisses as a way of establishing or maintaining dominance hierarchies.

Evidence for any of the hypotheses is rather slim. A study attempted to evaluate each of these hypotheses with Gunnison's prairie dogs (Creef 1993). For the Food Information Hypothesis, a colony of marked individuals was presented with three different novel food items—grapes, almonds, and horse pellets—and the frequency of greet-kissing after the introduction of the food items was compared with the frequency of greet-kissing during a control period prior to the appearance of the novel food items. With the introduction of each food item, the frequency of greet-kissing decreased, as the animals spent more time foraging for the novel items, suggesting that perhaps the information about the food items was transferred to other animals, encouraging them to forage more. For the Individual Recognition Hypothesis, the study observed the aggressive interactions of individuals within two separate territories, making the prediction that if the greet-kiss is necessary for individual recognition, then most aggressive interactions would be preceded by a greet-kiss. There was no evidence that aggressive interactions followed a greet-kiss, which might be expected if the animals were using vision to recognize other individuals. For the Dominance Maintenance Hypothesis, observations were made to determine which gender/age class tended to be more dominant over others, as defined by initiating a chase, winning a fight, or displacing the other individual. Adult males generally were more dominant over females and juveniles by chasing, displacing, or by winning fights. Both the adult females and the juveniles tended to initiate greet-kisses with adult males, but there was no difference in who initiated greet-kisses among the adult females and the juveniles. Among adult males, greet-kisses between males from different territories always were followed by a chase or a fight, while greet-kisses between females of the same territory, or between adults and juveniles within a territory, were usually amicable, with each individual foraging or standing next to the other following a greet-kiss. Thus, the picture is still not clear about the functions of a greet-kiss. Perhaps the behavior originated for any of the reasons given by the above hypotheses, and has persisted as a mechanism of social reassurance, somewhat like a human kiss.

Mutual Grooming

In addition to greet-kisses, black-tailed prairie dogs engage in another social behavior, mutual grooming (often called allogrooming) (Hoogland 1995). This behavior is similar to the grooming of primates. One prairie dog will comb through the fur of another, looking for parasites such as fleas. Since prairie dogs are often infested with fleas, this behavior can be beneficial in reducing the numbers of fleas on an animal. Unlike the black-tailed prairie dogs, the Gunnison's, white-tailed, and Utah prairie dogs (the white-tailed group) do not engage in mutual grooming, instead grooming only themselves (called autogrooming). The lack of mutual grooming has been one reason why the white-tailed group has been considered to be less social than the black-tailed prairie dogs. Not much is known about the behavior of Mexican prairie dogs, but being close relatives of the black-tails, we can predict they also have mutual grooming.

Infanticide

Another social behavior of black-tailed prairie dogs that is not shared by the other species is infanticide, with females killing the young of other mothers. Infanticide is relatively common among mammals, particularly when females are stressed due to excessive crowding or environmental disturbances (Hrdy and Hausfater 1984). Lactating black-tailed mothers kill the offspring of other mothers, often those of their close kin (Hoogland 1985, 1995). Females run into the burrows of other mothers when only the young are there and either kill the young underground, or drag the young onto the surface and kill them there. After killing the young, a female usually emerges onto the surface with a bloody mouth, which she wipes on the ground before she returns back to her own burrow. The evidence for infanticide is based on observations of young being cannibalized on the surface, observations of females cleaning their bloody mouths, and excavations of burrows where the infanticide was presumed to take place. Based on this evidence, there were 65 cases of infanticide observed over a four-year period out of 591 litters, from 1984 to 1988, at a colony at Wind Cave National Park, South Dakota. The female committing the infanticide was usually a relative of the mother whose young were killed—often half-sisters or full-sisters. She was also a female who was lactating (78% of the killer females), or had recently lost her own litter (12%), or had juveniles who had emerged from their burrow (5%), or was pregnant (3%). Only 2% of the females who killed other babies were not involved with their own pregnancy or rearing of young. The killer females were chased away by the mother if she detected them, either before they attempted to enter her burrow or after they killed her litter. However, within days of having their litter killed, the mothers went back to amicable behaviors such as greet-kissing with the females who killed their offspring (Hoogland 1995).

Once the young emerge above ground, the black-tailed females engage in communal nursing (Hoogland 1995). At Wind Cave National Park, between 48% and 62% of the pups within coteries nursed communally during 1986–87 (Hoogland et al. 1989; Hoogland 1995). Also, up to 50% of the lactating mothers did not discriminate between their own young and the young of other females when the young tried to suckle (Hoogland 1995). To investigate this, Hoogland and his colleagues injected one lactating mother per coterie with 15 microcuries of a radionucleotide, waited one to three days, and then took samples of blood and scat of the juveniles in the coterie. As predicted by the communal nursing hypothesis, the radionucleotides showed up in most or all of the juveniles in the coterie, and not just in the juveniles belonging to the injected mother (Hoogland et al. 1989). None of the other species is known to nurse communally, although the Mexican prairie dogs, being closely related to the black-tails, might also have communal nursing.

Prairie Dog Personalities

Because prairie dogs have such a complex social life, an intriguing question is whether there is any scientific evidence that each prairie dog might have his/her own personality. After all, the social life of humans would be dull and boring if it were not for differences in personality. Can it be that prairie dogs also have personality differences? Anyone who has worked with prairie dogs knows that there are differences. For example, some animals alarm-call all the time (the Worried Willies and Nervous Nellies), while other animals hardly ever call. When being marked with fur dye for scientific study, some animals relax and wait for their release, while others struggle and try to escape. But all of this falls into the category of anecdotal evidence, the anathema of science—if it's not the product of an experiment or observations with rigid protocols and sample sizes, then it just doesn't make the grade as far as most scientists are concerned. But, it turns out there just might be some evidence suggesting personalities. One study looked at the way that black-tailed prairie dog pups use their time—how much time each pup spent feeding, scanning for predators, standing on its hind legs, playing, grooming, and vocalizing—from the time they emerged in late May until 10 weeks later in mid-August (Loughry and Lazari 1994). Each pup used its time in individually-distinct ways, suggesting that each pup had its own personality. This is further corroborated by studies in which prairie dogs discriminate between different species of snakes: not all prairie dogs in a particular test situation participate in harassing the snakes that are offered to them by experimenters (Loughry 1987a, 1987b, 1988, 1989; Owings and Loughry 1985). As with humans, some prairie dogs are bold and some are timid.

Interlude:
Social Behavior—Trapping Prairie Dogs

July 16. Today we are trapping prairie dogs to mark them so that we can keep track of individual animals. The traps are Tomahawk live traps, wire mesh cages about 2 feet long and 10 inches high (61 cm × 25 cm) with a trap door (Figure I1). We set them up at night, bait them with sunflower seeds, and then we check them first thing in the morning. If a prairie dog is in a trap we put a pillow case over the door, open it, and let the prairie dog run into the pillow case. We pick up the pillow case, and weigh the prairie dog. Then, like wranglers at a miniature rodeo, we coax the prairie dog's nose into the corner of the pillow case so that we can keep it still as we grab hold of it firmly with leather-gloved hands and bring it out. While holding the animal, we record its age and sex, then we paint a number on its sides with black hair dye, give it an ear band (or check and record its ear band if it already has one) or inject a microchip with an identification number. The whole process takes about three minutes for each prairie dog.

The first prairie dog of the morning is huge and active. He looks as if he could burst out of the trap at any minute but when we get him into the pillow case and grab him firmly he settles and is still until we let him go. His eyes are like brown shiny beads and his coat has a gray dusting to it, making him seem lighter than the dust on the ground. He has hardly any fleas; some prairie dogs are crawling with them. We give every prairie dog a dusting of flea powder each time they are caught, both for their benefit and ours. We have caught this one before. We record his ear tag number, retouch his number, 8A, with black hair dye and then we stoop down to let him go. At the point of release, we draw back our hands quickly, which is fortunate because we remember that 8A has a nasty habit of snapping his head around at the last minute to see if he can give us a good nip before he runs away.

Figure 11. Gunnison's prairie dog in a squirrel-sized live trap. (Photo by C. N. Slobodchikoff.)

We look at the next trap and see, to our amusement, an old female, fat from our sunflower seeds, who we have trapped every morning this summer. We have records on this female for several years. She seems to have some sort of seniority in the colony that enables her to cross all territory boundaries with impunity. In an attempt to quit catching her, we stopped placing traps in the corner of the colony where she resides. But, regardless of where we place the traps, she ends up caught in one almost every morning, surprising us with her new location. We quickly let her go without picking her up and move onto the next prairie dog as the morning continues.

At the beginning of our project, it never occurs to most of our helpers that prairie dogs could be so different from each other. Because most prairie dogs look remarkably alike from a distance, our helpers never imagine that they would be able to recognize certain prairie dogs so easily up close during our trapping sessions. All of the animals look so nondescript at first in their small brown bodies. As scientists we have been trained to look at these animals as our "subjects" and our "samples," and to not project feelings or personalities or uniqueness on them. Naming them with numbers is one way to avoid this pitfall. It is a lot easier to anthropomorphize a "Joey" and "Sally" than an 8A or 9L. However, as scientists we are also trained to be observant, and so, as we work closely with these "subjects," we cannot escape the fact that we are working with individuals that both look and act differently from each other.

Prairie dogs can be very different physically at close range. Some are big and hefty, some have longer legs and torsos, some have pointier heads, thicker whiskers, and rusty flecks in their coats. But, it doesn't stop there because their actions are different too. For example, the way a prairie dog handles being held can be different. 7C's small foot grips against our glove and holds tight, nails almost piercing the leather. 3A squeezes its claws slowly and then releases them and then repeats these actions for the duration of being held. 5M kneads and scrapes the glove, while 2R keeps completely still. The way each looks at us is different too. Some watch us the whole time, making eye contact when we look at them. Others will watch the colony or some distant point. 6U always looks through half closed eyes with an unfocused gaze until the moment of release when her eyes snap wide open as she runs off.

Prairie dogs can differ in their behaviors too. Some males are very aggressive and immediately attack other animals who stray into their territory. Other males are much more tolerant. They will watch the intruding male and wait to attack him when it is clear that the intruder has no immediate intention of leaving. Some females spend a considerable amount of time greet-kissing other females and their young, while others spend relatively little time greet-kissing. Some mothers are very solicitous of their young, and watch carefully to make sure that the young pups do not stray too far from their burrows. Other mothers do not seem to care very much about what happens to their young, and are more concerned with socializing with other adult females in the territorial group. A few animals are always alarm calling, the "Worried Willies" and "Nervous Nellies," even when there is nothing out there to elicit any alarm calls. Other animals ignore these constant callers.

We don't need to anthropomorphize prairie dogs to be amazed at the diversity of physical traits and behavioral responses that are apparent among these animals. Each prairie dog embodies a unique mix of physical and behavioral traits like no other individual. It is a plain and simple fact that stares us in the face. As scientists, we are often trained to think more about the level of the population, species, and genus or even higher to ecosystem dynamics, ecotones, life zones, and biomes. However, when thinking about humans, we remember the importance of the individual person, because that is what we experience on a daily basis. As trained wildlife biologists, ecologists, or environmental scientists, we forget too easily that animal populations and species are made up of individuals, and that each individual is distinct and important whether it is a wolf, monkey, elephant, bird, or prairie dog.

In Ray Bradbury's short story, "The Sound of Thunder," a time traveler hunting dinosaurs in the Cretaceous period steps on a primitive butterfly and the whole future of the Earth, and the species that evolve to inhabit it, changes completely. It is a bit dramatic to suggest that a time traveler similarly ending the life of a certain key prairie dog would have a drastic effect on the future of our world. Nevertheless, when we are holding these prairie dogs in our hands, and seeing their individual differences, we are reminded that each prairie dog's actions affect the trajectory of life on the colony: whether they breed and who they breed with, at what point in their life they meet their demise, what they eat and when, if they happen to alarm call or not. Each prairie dog is an active and contributing participant in determining the future of a colony, a population, a species, a genus, an ecosystem, and perhaps even our living planet and each is going about it in its own way.

4

Communication

All forms of communication are tied closely to the senses. They can be tactile, olfactory, visual, vocal, magnetic, or electrical. For example, honeybees perform intricate dances on the inside of the hive, a form of communication to other hive members detailing where different flower patches can be found. Anyone who has ever dropped baking soda on an ant trail knows that ants communicate primarily with chemical signals that are transferred through an olfactory system, and anyone who has ridden a horse has experienced tactile communication firsthand. The world is full of signals of all different varieties and sorts spinning through the air, whisking through the water, brushing the skins and tickling the ears of a plethora of different beings. Communication happens at the cellular level through chemical and electrical impulses. It occurs on the genetic level with immense amounts of information packed on DNA, folded into chromosomes that form life's blueprint, and also incorporated in RNA. It happens in mating rituals, in hunting forays, and even between entirely different species.

Like most other animals, including humans, prairie dogs are extremely dependent on a communication system for survival. Prairie dogs have an extensive communication system, using several different kinds of signals. One of the principal forms of communication is through the use of sounds. All the species of prairie dogs produce alarm vocalizations when they see a predator, and at least some of the species have a variety of other vocalizations that are not associated with predators (Waring 1966, 1970). Another form of communication is through the use of visual signals. These are in the form of wagging their tails (tail-flagging), and also possibly standing upright in an alert posture (posting) (Owings and Hennessy 1984). A third form of communication is through the use of olfactory signals and odor cues (Halpin

1984). Of the three different kinds of signals, the acoustic or sound signals have been studied the most intensively.

In this chapter, we show that the communication system of prairie dogs is very complex. The alarm calls of Gunnison's prairie dogs can convey a considerable amount of information about the type of predator and something of the physical attributes of the predator. In addition, Gunnison's prairie dogs have social chatters that have a complex acoustic structure. Black-tailed prairie dogs have a "jump-yip" vocalization that appears to be lacking in the other four species.

Alarm Calls

The primary acoustic signal that prairie dogs produce is the alarm vocalization, a loud, often repetitive call that sounds somewhat like a bird chip. The sound is often called an alarm bark, because early Anglo settlers on the Great Plains of the Midwest thought that the alarm calls of the prairie dogs sounded like the barks of distant dogs (this is probably also the origin of the name, "prairie dog"). The alarm call is given by one or more prairie dogs within a colony when detecting a predator, and produces escape responses on the part of other prairie dogs hearing the call. All of the five species of prairie dogs produce alarm calls, but the acoustic structure of the calls differs among the different species.

A traditional view of alarm calls is that they function primarily to warn relatives of impending danger (Hoogland 1983, 1996). In experiments at a colony of black-tailed prairie dogs at Wind Cave National Park, South Dakota, Hoogland (1983) dragged a stuffed badger *(Taxidea taxis)* through the colony and observed the alarm-calling patterns of the prairie dogs in response. He found that although both the males and the females produced alarm calls, almost twice as many females (22 females) called as males (13 males). Also, twice as many females (11 females) called more often after their young emerged as males who called after the young appeared above ground (5 males). Although these numbers represent a small sample size, Hoogland (1983) concluded that females call more often than males to warn their close genetic relatives, partially because 5 of the males who called decreased their frequency of alarm-calling when they moved to other coteries where they did not have close genetic relatives. Using similar techniques of dragging two stuffed badgers through a colony of Gunnison's prairie dogs at Petrified Forest National Park in Arizona, Hoogland (1996) found that the alarm-calling patterns were variable. No animal called in 33% of the experimental trials (41 out of a total of 126 trials), while at least one prairie dog called in 67% of the trials (85 out of 126). Of the 50 alarm-calling females that Hoogland (1996) observed, 35 called when their offspring were present, a statistically-significant number. Among the 25 alarm-calling males, there was no significant difference for the males calling when kin or non-kin were present. These results are consistent with the hypothesis that prairie dogs call to warn relatives, because DNA analyses have shown that females mate with multiple males, many of them from different territories (Travis et al. 1996; Haynie et al. 2003), and the offspring of the males are likely to be scattered throughout the colony, while the offspring of the females are likely to be concentrated within their natal territories. Also, among the Gunnison's prairie dogs, there is a high level of relatedness among all of the animals within a colony (Travis et al. 1997) (See chapter 3).

We know more about the alarm calls than we know about the other forms of communication. Much of this comes from studies of the alarm call system of the Gunnison's prairie dog. However, the other species probably also have an alarm call system that is similar to that of the Gunnison's prairie dogs, although the details of the acoustic structure of the calls are different.

The alarm call system of the Gunnison's prairie dog is a very complex one. Gunnison's prairie dogs have different alarm calls for several different species of predator (Placer and Slobodchikoff 2000, 2001): humans, red-tailed hawks *(Buteo jamaicensis)*, coyotes *(Canis latrans)*, and domestic dogs *(Canis familiaris)* (Figure 4.1). Although some people are surprised that prairie dogs have a distinct call for humans, it is really quite understandable in an evolutionary sense. Humans have been preying on prairie dogs for hundreds, if not thousands, of years. For the Gunnison's prairie dog, both the Hopi and the Navajo, who live in the same areas where these prairie dogs are found, have recipes for cooking the animals (Slobodchikoff et al. 1991). Many Native Americans throughout the range of prairie dogs still eat them. Prairie dog bones have been found in some of the prehistoric Anasazi settlements, suggesting that humans have been catching and eating prairie dogs for a long time. At the present time, humans regularly shoot prairie dogs for a variety of reasons: target practice, recreation, or population control.

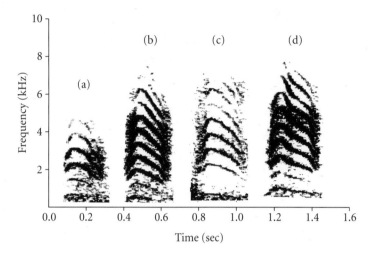

Figure 4.1. Gunnison's prairie dog alarm call sonograms for four different predators.
(a) Hawk. (b) Human. (c) Coyote. (d) Domestic dog.

When a human approaches a Gunnison's prairie dog colony, the first prairie dog to detect the person gives an alarm call. This alarm call can be in the form of a single sharp bark, if the human is approaching at a very fast speed, such as running, or it can be in the form of a series of repetitive barks, if the human is approaches slowly by walking (Kiriazis 1991). All the other prairie dogs in the vicinity of the human immediately run to their burrows and dive inside, so that within seconds of the alarm call there is not a prairie dog that is visible to the human. Many of the prairie dogs do not retreat very far into their burrows, but sit inside the opening and poke their heads out far enough so that they can watch where the human is going. If the human starts to head toward them, they retreat farther into their burrows (Kiriazis and Slobodchikoff 2006).

When a red-tailed hawk dives down toward the prairie dog colony, the first animal that detects the hawk gives a single sharp bark, and all the prairie dogs in the immediate flight path of the hawk run to their burrows. The animals that are outside the immediate flight path stand upright and watch the progress of the hawk. Unlike the alarm calls elicited by a human, repetitive alarm barks are not used, probably because hawks dive at a very fast speed and therefore have a better chance of locating and catching a repeated alarm caller than a human would (Kiriazis and Slobodchikoff 2006).

The response of Gunnison's prairie dogs to a coyote is very different from the response to either humans or to hawks. When a coyote approaches the colony, many animals start to produce an alarm call. Alarm calls can be heard from throughout the colony. Upon hearing an alarm call for a coyote, the prairie dogs run to the lips of their burrows and stand upright on their hind legs, watching the progress of the coyote. Other animals that were below ground emerge and also stand at the lips of their burrows (Kiriazis and Slobodchikoff 2006). If the coyote is moving slowly, the chorus of alarm barks is repeated relatively slowly, but as the coyote speeds up and starts to run, the interval between each alarm bark shortens in proportion to the speed of the coyote (Kiriazis 1991).

When a domestic dog approaches the Gunnison's prairie dog colony, the response is somewhat similar to that for a coyote. There is multiple chorusing of alarm calls from several individuals simultaneously. However, unlike for the coyote, the animals that hear a domestic dog alarm call do not run to the lips of their burrows—instead, they stand upright on their hind legs wherever they have been foraging, and watch the progress of the dog through the colony. When they decide that the dog has come too close to them, only then do they run to their burrow. But, as with the coyote, other animals that are belowground emerge and stand at their burrow openings, watching the progress of the dog (Kiriazis and Slobodchikoff 2006).

An important point to note is that there is a qualitative difference between the responses (Figure 4.2). The human elicits a colony-wide running to the burrows and diving inside. The hawk elicits running only in its immediate flight path. The coyote elicits a running to the burrows, standing at the lip of the burrows, and belowground animals emerging. The domestic dog elicits standing in place, and running to the burrows only when the dog has gotten too close. Previously, some authors had suggested that all ground squirrels, including prairie dogs, have essentially the same response to all predators, running to their burrows and going inside (Macedonia and Evans 1994; Evans 1997) and that the escape behaviors of primates (such as vervet monkeys, *Cercopithecus aethiops*) are more complex, because the primates have several different escape responses, rather than the single one of the ground squirrels. However, the Gunnison's prairie dogs have as much if not more complexity in their escape responses than the vervet monkeys, who have three types of calls for different predators and three different types of escape responses

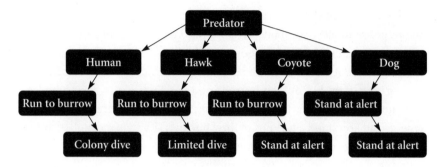

Figure 4.2. Escape responses of Gunnison's prairie dogs when confronted with different
predators. Although for humans, hawks, and coyotes, the initial response of
the prairie dogs is to run to their burrows, the subsequent response differs.
For a human, there is a colonywide retreat into the burrows, while for a hawk,
only the prairie dogs in the immediate flight path of a diving hawk go into
their burrows. For a coyote, the prairie dogs stand at alert at the lips of their
burrows, and for a domestic dog, the prairie dogs stand at alert wherever they
are foraging.

(Cheney and Seyfarth 1990). Vervet monkeys have acoustically different
alarm calls to three different predators: leopards *(Panthera pardus)*, martial
eagles *(Polemaetus bellicosus)*, and pythons *(Python sebae)*. In each case the
escape behavior is different. For a leopard call, the monkeys run up into
acacia trees and into the periphery of the branches, while for an eagle call
the monkeys also run up into a tree, but stay near the trunk where they are
more inaccessible to the eagle. For a python call, the monkeys look along the
surface of the ground (Cheney and Seyfarth 1990). Unlike the prairie dogs,
who can distinguish between similar-appearing predators such as coyotes and
domestic dogs, the vervets generalize their calls to include other predators as
well. Leopard calls are also given for caracals *(Felis caracal)* and servals *(Felis
serval)*, martial eagle calls are also given for crowned eagles *(Stephanoaetus
coronatus)*, and python calls are also given for mambas *(Denroaspis* spp.) and
cobras *(Naja* spp.) (Cheney and Seyfarth 1990).

Although the prairie dogs have different alarm calls for different predators and have different escape behaviors for each predator, we still must have
evidence that the calls themselves produce the appropriate escape behaviors
in the absence of a predator. An alternative explanation might be that the
calls serve to merely warn other animals that there is a predator approaching,
and the animals then adopt an appropriate escape behavior upon seeing and

identifying the predator. Potentially, the calls, although different, could have no meaning beyond an arousal function. In the case of the vervet monkeys, Cheney and Seyfarth (1990) used playbacks of predator calls when no predator was present, and found that the monkeys used the same escape behavior on hearing the call as they used when the predator was actually present.

To test whether the different alarm calls are meaningful to prairie dogs and produce the appropriate escape behaviors, Kiriazis and Slobodchikoff (2006) did a playback experiment. Studying the animals at two colonies in the vicinity of Flagstaff, Arizona, they first recorded the alarm calls and videotaped the escape behaviors of prairie dogs to naturally-occurring coyotes, domestic dogs, red-tailed hawks, and humans. Then they played back the alarm calls to the prairie dogs when no predators were present, and videotaped the escape responses. As a control, they broadcast a nonsense sound that spanned the same frequency range as the alarm calls, to determine if the prairie dogs might be responding to just anything that resembled an alarm call. The nonsense sound produced no response in the prairie dogs. The animals merely continued to forage or carry on with their normal activities. Broadcasts of alarm calls for predators, however, produced immediate escape behaviors. For each of four types of predator call, coyote, domestic dog, red-tailed hawk, and human, Kiriazis and Slobodchikoff (2006) compared the escape responses of the prairie dogs elicited by the predator's presence and the escape responses elicited by just the alarm calls with no predator in view. The differences between the escape responses triggered by alarm calls only and the escape responses elicited by the actual predator were not statistically significant for all four types of calls. This demonstrated that information about the identity of the predator was encoded in the alarm calls, and that prairie dogs did not need to see the predator in order to adopt the appropriate escape strategy.

In addition to distinct alarm calls for coyote, domestic dog, red-tailed hawk, and human there appear to be calls for other species as well. There are different calls for a badger *(Taxidea taxus)*, a grey fox *(Urocyon cinereoargenteus),* and a cat *(Felis cattus).* There are also calls for non-predators as well, such as acoustically different calls for a pronghorn antelope, an elk, and a cow. The latter calls raise the question of whether these are alarm calls, since none of these animals can eat a prairie dog. Of course, we can argue that perhaps all of these animals can startle prairie dogs the same way that a predator can startle prairie dogs, but the different escape responses to the different predators suggest that prairie dogs have good discriminatory abilities and would not simply be startled by common non-predatory animals in their environment (Slobodchikoff et al., in prep).

Perhaps, rather than being alarm calls, these calls communicate information to other prairie dogs about what is happening in their general vicinity. In this sense, the calls can function as a language, rather than as a response to an alarming stimulus. One experiment with Gunnison's prairie dogs suggests that there is a larger element of cognition in the vocal responses of prairie dogs than has previously been assumed, and that some elements of language might be involved. In this experiment, a colony of prairie dogs were presented with three different kinds of silhouettes, all made out of plywood and painted black: a coyote, a skunk (*Mephitis* spp.), and an oval. The coyote and skunk silhouettes were life-sized, and the oval was the same size as the coyote. All of the silhouettes were mounted on pulleys and were drawn by rope across the colony from a hiding place on the edge of the colony. The calls of the first animal to respond to each presentation were recorded and subsequently analyzed. All of the prairie dogs were marked with distinctive fur-dye marks, so the identity of each individual caller was known (Ackers and Slobodchikoff 1999).

The results of this experiment showed several things. One was that all of the animals called in the same way to the coyote silhouette, in a different way to the skunk silhouette, and in still a different way to the oval silhouette. Although there were some individual differences between animals, in the same way that there are individual differences in human voices, there was still remarkable similarity in the calls of the animals for each of the different silhouettes. The calls for the coyote silhouette were fairly similar to the calls for a real coyote, with a few differences that seemed to be related to the structure of the coyote silhouette. This indicates that each animal calls in a consistent way for each of the different types of predators—that the differences that were observed previously are not due to random chance differences. For example, if the differences were due to random chance, let us suppose that animal A always calls in only one certain way, and animal B always calls in a different way, regardless of the species of predator. Now let us suppose that we do an experiment where we have a human walking through the colony, and we happen to record only animal A. In another experiment where we have a coyote walking through the colony, we happen to record only animal B. We might then think that there was a different call for humans and coyotes,

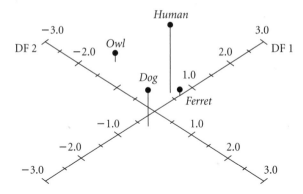

Figure 4.3. Discriminant function analysis (DFA) for alarm calls produced by six different prairie dogs in a laboratory setting, elicited by four different potential predators: human, domestic dog, great-horned owl, and European ferret. None of the prairie dogs had ever seen either a great-horned owl or a European ferret previously, but all of the prairie dogs consistently had different calls for these two predators, as well as different calls for the human and the dog.

whereas in fact animal A always calls the same way for any kind of predator, just as animal B always calls in a different way for any predator. The silhouette experiment conclusively showed that all the animals call in the same way for a coyote silhouette, in a different way for the skunk silhouette, and in a still different way for the oval silhouette.

Another experiment with novel stimuli was done in the laboratory with six captive prairie dogs (Slobodchikoff, unpublished data). The experiment involved placing each prairie dog by itself into an acoustically-isolated room, and introducing at random one of four possible predators: a human, a domestic dog, a great-horned owl, and a European ferret. Because owls are nocturnal and prairie dogs are diurnal, it is likely that none of the prairie dogs would have seen an owl before, nor would they have seen a European ferret because these ferrets do not occur in the wild. Each of the prairie dogs had a consistent call for a human and a different one for the dog, but they also had another call for the great-horned owl and still a different call for the European ferret (Figure 4.3). Analysis of the resulting calls with Discriminant Function Analysis (DFA), a statistical program that classifies calls on the basis of how variable they are, showed that each of the calls for the four types of predators was statistically significantly different.

These two experiments showed that prairie dogs can describe objects that are completely novel to them. The oval silhouette was something that the prairie dogs had never seen before, and yet they all consistently had the same kind of calls in response to the appearance of this silhouette. This suggests that they can modify their alarm calls according to the structural features of the object that they see, and it also suggests that they have the capability to produce alarm calls in response to novel objects in their environment that they have never previously encountered. Perhaps prairie dogs have some kind of neural template in their brains that contains a variety of descriptive elements, and when they encounter an animal in their environment, they can combine these elements in different ways to describe this animal, much in the same way that vocabulary words can be used in a language (Slobodchikoff 2002).

The acoustic structure of the calls can also describe the physical features of individual predators. Within a category of predator, such as human, Gunnison's prairie dogs can modify the structure of the call to encode information about the general size, shape, and color of clothes that the human is wearing. In a series of experiments to demonstrate this, Slobodchikoff et al. (1991) asked four humans to walk separately through two prairie dog colonies near Flagstaff, Arizona, wearing different clothes. In one set of walk-throughs, all of the humans wore the same white laboratory coat, which hung loosely and obscured the shape of each human. In another set of walk-throughs, each human wore a different-colored t-shirt: blue, green, yellow, or gray. In a third set of walk-throughs, two humans alternated wearing a yellow and a white t-shirt. In the white-laboratory-coat experiment, the prairie dogs did not have any significant differences in their calls for two of the humans who were similar in size and shape, but had a significant difference between two of the humans who differed in their slenderness, and had a significant difference in all of their calls for the fourth human who was shorter than the others. Apparently, the white laboratory coat obscured the details of the physical features of some of the humans. In the four-colored-T-shirt experiment, the prairie dogs had significantly different calls for all four of the humans, and an analysis of calls showed that one component coded for colors of the shirts. In the experiment in which two humans traded shirts, the prairie dogs had significantly different calls for the two humans, and one component of the calls coded for either white or yellow.

This means that if the human is tall, thin, and wearing a blue shirt, the alarm

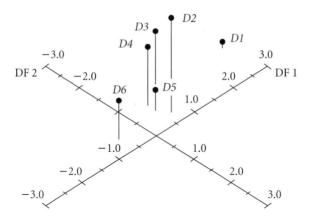

Figure 4.4. Discriminant function analysis (DFA) for alarm calls produced by six different prairie dogs in a laboratory setting, elicited by six different domestic dogs of differing size and coloration. D1 was a black dog, and D6 was a white dog, while D2–D5 were dogs with yellow or brown coats. The discriminant function axis DF1 correctly placed the different dogs according to their coat colors.

call would be distinctly for a human, but would have structural elements that code for size and shirt color. The same human wearing a green shirt would elicit a slightly different call, still distinctly a human call, but with structural modifications that encode for a green shirt. Similar differences can be found for the size and shape and coat color of coyotes and domestic dogs.

In another experiment performed in the laboratory with the six captive Gunnison's prairie dogs (Slobodchikoff, unpublished data), the animals were shown six different domestic dogs, varying in coat color from pure black to pure white. The methodology was the same as that described above for the four different predators. Each prairie dog had consistently different alarm calls for each of the dogs, and Discriminant Function Analysis showed that these differences contained information about the coat colors of the dogs (Figure 4.4).

Although the most experimental work in this regard has been done with Gunnison's prairie dogs, a recent study of the response of the other four species of prairie dogs to a human wearing a green shirt versus the same human wearing a yellow shirt has shown that all of the prairie dog species can incorporate this kind of descriptive information into their alarm calls (Frederiksen 2005). However, the exact acoustical structures, in terms of elements of sound frequency and time, differ among the species, suggesting

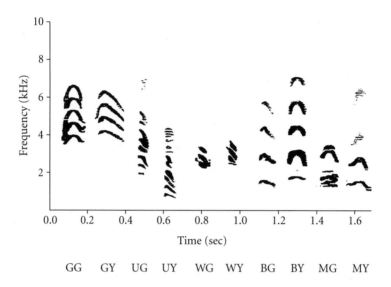

Figure 4.5. A comparison of the calls of all five species of prairie dogs elicited by the same human wearing either green (G) or yellow (Y) for Gunnison's (G), Utah (U), white-tailed (W), black-tailed (B), and Mexican (M) prairie dogs. For example, GG is a Gunnison's call for a green shirt, and GY is a Gunnison's call for a yellow shirt. (From Frederiksen 2005, with permission of the author.)

that a given prairie dog species would not be able to understand the communication calls of any of the other species (Figure 4.5).

Although it might seem surprising that the prairie dogs can encode such detailed information into their alarm calls, perhaps this has some adaptive value in terms of identifying individual predators. The colonies are spatially fixed, and the same individual predators keep coming back day after day. These predators might have individually-different hunting styles. For example, some coyotes hunt prairie dogs by walking through the colony, and rushing at any animals that appear to be slightly farther from their burrows than the other animals along their path. Other coyotes hunt by identifying a concentration of prairie dogs at a burrow, going up to that burrow, lying down next to the burrow, and waiting for up to an hour for a prairie dog to emerge. If a prairie dog does emerge, they immediately pounce, grab the prairie dog, give it a shake, and proceed with eating it. So, for the prairie dogs it might be important to know which individual coyote is walking through their colony. For example, they need to know whether they have to be careful about coming out of their burrows, or whether they simply have to make sure that they are right next to a burrow.

Along with descriptive information, Gunnison's prairie dogs can apparently incorporate some information about the potential dangerousness-level of the predator. This was demonstrated by field experiments (Long 1998), in which the same person initially appeared to the prairie dogs over the course of two weeks in two different contexts: one in which he wore a white laboratory coat, sunglasses, and walked slowly in a prescribed path through the colony; the other in which he wore a gray laboratory coat, carried a simulated rifle, walked an erratic path and occasionally stopped to lunge at any prairie dogs whose heads were peeking up out of their burrows. The white vs. gray laboratory coats were the only difference in physical features, since the person was the same in each case. After the training period, the person randomly appeared either in the white or the gray laboratory coat, each time walking slowly in a prescribed path without lunging at any of the prairie dogs. The prairie dogs responded with more alarm to the gray coat instances, and had a different alarm call for the gray coat appearances than they did for the white coat appearances, suggesting that they learned and remembered that the gray coat situation was potentially more threatening than the white coat situation.

Black-tailed prairie dogs appear to have a similar system of descriptive elements in their alarm calls. In a series of experiments with a colony of black-tailed prairie dogs in Texas, repeating the design of the experiments done with Gunnison's, the black-tailed prairie dogs had a human call into which they incorporated information about the size, shape, and color of clothes that the humans were wearing (Frederiksen and Slobodchikoff 2007). In addition, the black-tails were exposed to another test: one of the human volunteers fired a shotgun after the initial prairie dog calls were recorded in response to this person without a gun. After he fired the shotgun, two things happened. One was that the prairie dogs modified their calls to incorporate a different element, apparently associated with a gun. The other was that as soon as this person appeared, even when he was not carrying a gun, the prairie dogs all quickly disappeared, making it difficult to get a large sample size of recordings for this person.

Another feature of the alarm calls of Gunnison's prairie dogs is that there are both regional and local dialects. At the local level, each colony has slightly different acoustic features in the alarm calls for each predator species. At the regional level, these differences increase with greater geographic distance between colonies. For example, the differences in acoustic structure between the alarm calls of colonies of Gunnison's prairie dogs in Colorado and Arizona are greater than the differences between colonies in Santa Fe and Taos, New Mexico. However, throughout the range of the Gunnison's prairie dogs, a call for a human is still recognizable as having the same basic structure, regardless of the dialect regions (Slobodchikoff and Coast 1980; Slobodchikoff et al. 1998). An experiment that still needs to be done is to play back the alarm calls of one colony to other colonies in the local region and throughout the geographical range, and to see at what point, if any, the alarm calls fail to elicit a response from the listening prairie dogs.

The dialects could be due to learning, genetic differences, or differences in the habitat structure of each colony. The genetic structure of colonies that are separated by a few kilometers is known to differ (Travis et al. 1997), but whether these genetic differences are related to differences in dialect is unknown. The habitat structure of colonies, and areas within colonies, differs (Slobodchikoff et al. 1988), and this can potentially affect the frequencies of sound that are transmitted through vegetation in a colony. A study with Gunnison's prairie dogs assessed the vegetation structure of colonies in the vicinity of Flagstaff, Arizona, and played back alarm calls from each of the colonies on all of the other colonies, to assess how sounds attenuate (Perla and Slobodchikoff 2002). The alarm calls from a colony attenuate the least (carry farther) on the colony where they were originally recorded, rather than on any of the other colonies. This result seems to be partially related to vegetation structure on each colony, in that sound frequency components, but not timing components, differ from one colony to another (Perla and Slobodchikoff 2002). It is also entirely possible that dialects could arise through learning; the vocalizations that carry sound more efficiently within the habitat structure of a particular colony are those that persist from one generation of prairie dogs to another.

Black-tailed prairie dogs have a vocal signal that is not found among the other prairie dogs—the jump-yip call (King 1955) (Figure 4.6). In this call, a prairie dog stands up on its hind legs, reaches upward with its front legs, and emits a "yip" vocalization. This vocalization appears to be contagious, in the sense that other animals also produce this same vocalization upon hear-

Figure 4.6. Jump-yip behavior of a black-tailed prairie dog. (Courtesy of Russell Graves.)

ing an initial jump-yip, and a wave of jump-yips can travel through all the prairie dogs in the immediate vicinity of the initial caller. The function of the jump-yip is not clear. Some possibilities are that it could function as a territorial advertisement, or as a social signal, or as a warning signal, or as an all-clear signal when a predator has gone (Hoogland 1995). Black-tailed prairie dogs will often give a jump-yip call in response to snakes, particularly to rattlesnakes and bullsnakes, and the acoustic structure of the jump-yips appears to vary according to the type and size of snake (Owings and Owings 1979; Owings and Loughry 1985). The acoustic structure of the jump-yip call

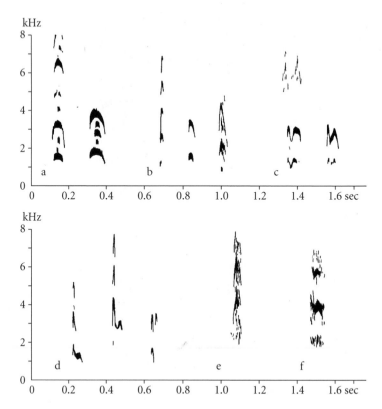

Figure 4.7. Different vocalizations of black-tailed prairie dogs. Letters designate different
     groups of vocalizations. (From Smith et al. 1977, with permission from
     Elsevier.)

has not been analyzed thoroughly, and it is possible that there will prove to be
different types of jump-yips associated with different contexts, just as there
are different types of Gunnison's alarm calls. Both the jump-yips and the
alarm barks of black-tailed prairie dogs show a considerable amount of variation (Smith et al. 1977; Owings and Morton 1998) (Figure 4.7), and further
analysis of the calls may reveal a complexity that is similar to that found in
the Gunnison's prairie dog calls.

    Both black-tailed and Gunnison's prairie dogs can become habituated to
the frequent presence of humans and cease responding with alarm calls as
rapidly as when they only see humans occasionally. In one study of black-
tailed prairie dogs living in city environments located in commercial areas
of Boulder, Colorado, the animals responded with alarm calls at shorter

distances to a human intruder than prairie dogs living in country areas 2–3 kilometers outside of Boulder (Adams et al. 1987). Response distances were approximately 9 meters (about 30 feet) in the city environments, and between 43–60 meters (about 130–180 feet) in the country environments. Once the animals disappeared in their burrows, the city animals came up again in a much shorter time span than the country animals. Another study of alarm calling at Wind Cave National Park, South Dakota, counted the number of alarm calls elicited from black-tailed prairie dogs at two sites, one a colony that was isolated from visits by tourists, and the other a colony that was visited frequently by humans (Motiff 1980). At both sites, the prairie dogs called more frequently the closer the intruding human was to their burrow, but the isolated colony had more frequent alarm calling than the colony that was visited by people, suggesting that the prairie dogs habituate to the presence of humans when they encounter them often. Gunnison's prairie dogs cease alarm calling and stop running to their burrows upon repeatedly seeing the same human appearing in the colony (Ackers 1997).

Other Vocalizations

Some of the prairie dog species, if not all, have other vocalizations that seem to occur in a social context (Hoogland 1995; Smith et al. 1977; Waring 1970). These vocalizations, which can best be called social chatters, have been studied extensively in Gunnison's prairie dogs, where the social chatters consist of at least nine different kinds of syllables (Gilbert-Parker 1995). These range from alarm-call like syllables, to low-pitched growls, to frequency-modulated sounds that have a complex structure. Some of the chatters seem to have specific behaviors associated with them. Group 3 and Group 8 calls (Figures 4.8, 4.9) are given primarily by a prairie dog standing in an aggressive posture; Group 8 calls are also given by a prairie dog sitting up on its haunches; Group 2 calls are given by an animal in an alert position, standing upright on its hind legs; Group 6 calls are given primarily by a running animal; Group 7 calls are given either by two males who are fighting or by juveniles who are play-fighting; Group 9 calls are given when the vocalizer is chasing another animal (Gilbert-Parker 1995). Other chatter calls have no obvious behaviors associated with them. Some of the chatter call groups can be combined into a string of syllables, producing a duet of one animal calling and another one

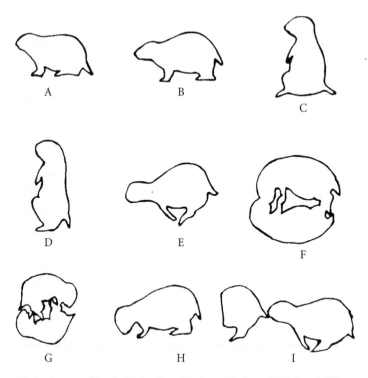

Figure 4.8. Postures used by chattering Gunnison's prairie dogs. (A) Relaxed. (B) Aggressive. (C) Semi-alert. (D) Alert. (E) Running. (F) Fighting. (G) Play-fighting between juveniles. (H) Foraging. (I) Chasing. (From Gilbert-Parker 1995, with permission of the author.)

responding. One animal will emit a string of syllables, and another animal will respond with its own string of syllables. Under these circumstances, nothing about the behavior of the animals that are calling changes, and it is difficult to decipher any possible meaning associated with this chatters, unlike the alarm calls where the presence of a predator and a distinct escape response have allowed the meaning of the different alarm calls to be deciphered. In colonies of Gunnison's prairie dogs, social chatters occur quite frequently. A common observation of people seeing prairie dog colonies for the first time is that the animals are very noisy.

Other vocalizations seem to occur in an aggressive context (Smith et al. 1977). Black-tailed prairie dogs have several different vocalizations. These

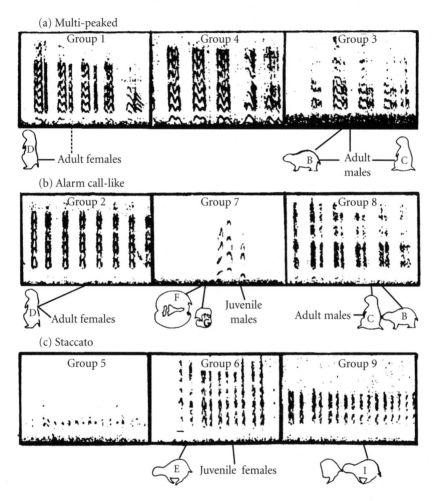

Figure 4.9. Types of chatters and associations between the chatter types, postures, and sex-age classes of Gunnison's prairie dogs. Solid lines indicate statistically significant relationships. (From Gilbert-Parker 1995, with permission of the author.)

include a rasp, which is a prolonged harsh vocalization used in the context of signaling an attack by a prairie dog during a fight; and a tooth-chatter, a low sound that prairie dogs make by chattering their teeth, primarily during disputes at the boundaries of territories. They also have a scream, given by animals that are being chased during a fight.

Visual Signals

Visual signals have been studied less intensively than acoustic signals in prairie dogs. Partly this might be because there is a relatively easy methodology available for studying acoustic signals, in that the signals can readily be recorded on tape and played back for sonographic analysis. Visual signals are much more difficult to quantify, even though they can be easily videotaped. Changes in posture or the angle at which a body part is held are difficult to measure with current techniques, and relatively difficult to quantify. However, it would be surprising if prairie dogs did not have a rich repertoire of visual signals that are given either concurrently with acoustic signals or that supplement the information contained in vocal calls. Black-tailed prairie dogs will raise their tails, spread out the fur on the tail, and wag the tail back and forth when they see a snake in a behavior called tail-flagging (Owings and Owings 1979). They will also do the same behavior when they are alarmed by a human (Frederiksen 1995). Gunnison's prairie dogs flag their tails as well. When a predator such as a coyote appears, Gunnison's prairie dogs stand upright on their hind legs in a behavior called posting. This upright stance allows them to see the approach of a predator better, but also makes them visible to other prairie dogs that might be foraging on all four legs. The number of animals engaging in posting behavior might be a signal to other prairie dogs that something is approaching the colony (Verdolin and Slobodchikoff 2002).

Olfactory Signals

Signals associated with odors are poorly understood. The animals have anal glands, and both black-tailed and Gunnison's prairie dogs have been observed dragging their rumps on the ground, perhaps as a way of leaving olfactory marks (Jones and Plakke 1981; Halpin 1984). Since other ground squirrels closely related to prairie dogs *(Marmota, Spermophilus)* have oral glands (Steiner 1974), it is likely that prairie dogs have them as well. Oral glands as well as saliva might function in the greet-kiss behavior of prairie dogs. This behavior consists of two animals coming together with their mouths open, pressing their tongues together, and standing with their mouths open for 1–2 seconds (King 1955; Steiner 1975) (see chapter 3). The greet-kiss behavior might function in allowing prairie dogs to identify individuals by their taste or smell, or it might function in disseminating information about the food that a prairie dog has been eating, or it might play a role in dominance interactions.

Prairie Dog Language

Prairie dogs have a highly complex system of communication including acoustical, visual, and olfactory signals. Most highly studied, because of the ease of quantifying different variables and the associated differences in response behaviors, are the alarm calls. Studying alarm calls of prairie dogs has significantly increased our understanding about the complexity inherent in non-human communication systems and has further narrowed the gap between human animals and other life forms.

The alarm calls of the Gunnison's prairie dogs have many language-like features. While language is difficult to define, there are some criteria, formulated by Charles Hockett in 1960, for the elements that should be included in a language. These criteria, often referred to as Hockett's Design Elements of Language (Hockett 1960), are as follows:

1. A vocal-auditory channel. Hockett suggested that any language should use sound, and be perceived by an animal's ears or other receptors that detect sound.
2. Broadcast transmission and directional reception. Messages are transmitted by broadcasting them, and animals that are listening can determine the direction from which the broadcast originates.
3. Rapid fading. Signals disappear quickly once they are broadcast, so that there is no confusion from echoes or other sources of sound.
4. Interchangeability. An animal can both broadcast a signal and receive the signal.
5. Total feedback. The animal that is broadcasting the signal can hear its own vocalization.
6. Specialization. The signals are designed to communicate information.
7. Semanticity. The elements of the communication system have semantic content, in that they refer to external objects and events.
8. Arbitrariness. The signals are arbitrary in their form, containing nothing that an external observer can immediately identify as an iconic property

of an external object. In English "bow-wow" would be an iconic property of a dog, because dogs bark with a sound that is like "bow-wow." The English word "dog" has both semantic content, meaning an animal that we think of as a dog, and is arbitrary, in that there is nothing in the word "dog" that would give someone a clue as to what we were talking about if she or he was not familiar with the word.

9. Discreteness. The signals are in discrete chunks, like words in English are discrete.

10. Displacement. The signals can refer to external objects or events that are distant from the animal in space or in time.

11. Productivity or Openness. New signals can arise to refer to new objects or new situations.

12. Duality. The signals are composed of smaller elements that combine to create larger messages. For example, English words are composed of phonemes, which have no meaning individually but can be combined with other phonemes into meaningful words.

13. Cultural or traditional transmission. The language can be learned and transmitted to others through learning.

The alarm calls exhibit almost all of these design features. They have design features 1–6, as do most acoustically based communication systems. They also contain semantic elements (design feature 7) coded in an arbitrary way (design feaure 8) that delineate the species of the predator, and provide descriptions of the color, size, shape, and dangerousness of the predator. The vocalizations have discreteness (design feature 9) in that they occur as discrete bursts of sound, although a single alarm call may be analogous to a sentence, because it contains information about the species of predator (noun-like elements), and a description of the individual features of the predator (adjective-like elements). The calls are repeated more rapidly when a predator is moving quickly, which provides verb-like elements, because the time of each call becomes shorter, referring to the speed of travel of the predator. The alarm calls show displacement (design feature 10) by referring to a predator that is somewhere distant from the prairie dogs, and productivity (design feature 11) by referring to novel objects that the prairie dogs have never seen before such as the ovals mentioned previously. Recent research into the structure of the alarm calls suggests that there may be duality (design feature 12) in that the alarm calls are made up of acoustic elements that are analogous to phonemes (Placer and Slobodchikoff 2004; Slobodchikoff and Placer 2006). Cultural transmission and learning (design feature 13) are unknowns at this point. It is entirely possible that parts of the vocalizations may be learned, but further experiments are needed to test whether or not learning plays a role in the development of the alarm calls.

Interlude:
Communication—Field Studies

May 28. We are setting up our gear for another recording session at the Snowbowl Colony near Flagstaff, Arizona. The colony is in part of a large wind-swept field that is about 2 miles long, surrounded by sparse clusters of houses on three sides and the major highway to the Grand Canyon on the fourth side. A small road leads off to the Snowbowl ski area near the top of the San Francisco Peaks, a name that we have adopted for our study colony. We have parked our cars at the side of the dirt road that represents the southern boundary of the field, which is a large swath of alpine grassland sitting in the middle of a ponderosa pine forest at an elevation of 7,200 feet (2,400 meters). The prairie dog colony here is small, about 1 hectare (2.5 acres). Another prairie dog colony can be seen in the distance, across the field in the northern part, also small, as many prairie dog colonies are these days.

We haul all of our recording gear out of the cars and check it for the second time to make sure that we have everything with us. The first time was in the laboratory, where we store all the gear when it is not being used. Everything seems to be in order. Two tape recorders, two directional microphones, a video camera, binoculars, notebooks, all are there. We check the tape recorders again, as we did an hour before in the lab. For reasons that we have never been able to understand, batteries and cables that worked fine when we checked them in the lab will sometimes fail out in the field, usually during the most critical recording moment. So we check and double check, and bring lots of spare parts.

With everything in order, we carry the gear to the barbed-wire fence that marks the edge of the field, and haul everything over to the observation tower that we have built to observe and record the prairie dogs (Figure I2). The observation tower stands about 10 feet (3 meters) high, made out of wood

Figure 12. Tower blind used for observing Gunnison's prairie dogs in the vicinity of Flagstaff, Arizona. (Photo by C. N. Slobodchikoff.)

and burlap. Four solid beams make up the legs, buried in the soil of the field. A plywood platform 6 feet (2 meters) up offers us a place to sit and store our recording equipment, and wooden railings around the platform help keep us inside and also offer a place to hang burlap sacks that hide us from the prairie dogs. We have cut holes in the burlap so that we can stick our microphones and binoculars out without the prairie dogs seeing our shapes on the platform. We hope that we are fooling the prairie dogs into not noticing us, and it seems to work, because once we are up there, the prairie dogs go about their business of foraging for food and pay no attention to us.

When we started building observation towers, we did not have a solid design or plan, relying on trial and error to build something that would support three people 6 feet above the ground. The first towers that we built were pretty rickety. Although 6 feet does not seem to be a very great height, it seems much higher when the tower begins to sway in the wind and the beams start to creak. A coincidence of events led to the design and building of a really solid tower. The first event was a mini-whirlwind, something relatively uncommon in northern Arizona, that picked up our tower and walked it a couple of hundred feet before dropping it on its side. Fortunately, no one was in the tower at the time. The second event was an engineer, who happened to be visiting at the time. He saw the fallen tower and proceeded to draw up plans for something that would withstand mini-whirlwinds and would be solid enough so that we would not have to risk motion-sickness when the wind was blowing. We built his tower, and it was solid.

Using a ladder nailed to one side of the tower, we climb up and lift up the equipment. It is now 5:00 AM, and the sun is rising. We make a point of getting to the observation tower early in the day, before the prairie dogs are up above ground, so that we minimize the disturbance that we cause by our arrival. As always, the wind is blowing, and the wind-chill is pretty intense. Although it is late May, we are sitting in our parkas, trying to stay warm. Once the sun comes out and drenches our tower with warm sunlight, we will have to shed our parkas, and the wind will help cool us off in the rising heat of the day.

Now we wait. Around 6:00 AM, the first prairie dogs start to emerge. Some sit up on their haunches by their burrow entrances and soak up the sun's rays. Others seem to be hungry, and immediately start to feed on grass and other vegetation. Over the next hour, more prairie dogs start to emerge, the late risers who probably slept in that morning, and soon the entire colony is awake and active. Some animals are feeding. Others are chasing each other, sometimes in play, other times in territorial disputes. We are all struck by how much activity goes on in a prairie dog colony, and how much fun it is to watch the animals scampering around.

We could watch them for hours, but our job this morning is to record alarm calls that the prairie dogs give to different predators. This means that we have to stay alert and watch the edges of the field to see if a coyote is coming, or scan the sky to see if there is a red-tailed hawk or golden eagle looking for a prairie dog meal. Because all of the prairie dogs are marked with black fur dye, with an individually-distinct combination of letters and numbers that tell us who each animal is, we also have to pay attention to where all of the animals are in the field so that we can see who calls first for a predator. Who calls first is important to us, because animals that call later, after another animal has called, could be responding to the alarm call and not the predator, and we need to be certain that we are recording a call that is given in response to the predator. We call tell where each animal is in the colony through a system of numbered stakes that we have placed at 10 meter (30 feet) intervals in a grid, so that we can locate each animal on a map that we have made of the area. We also need to stay alert so that we can record on video the escape responses of the prairie dog when they see the predator and when they hear an alarm call. Recording an alarm call given by the animal that first sees the predator means that we have to be able to point our directional microphone, which records sound only from the direction in which it is pointed, toward the animal that first makes the call. Between video camera, microphone, and recorder, plus watching the animals to identify individuals and note their positions in the colony, it is a full-time job for three people.

The morning stretches and the heat of the day sets in. No predators have come by, something that is unusual, because there are several coyotes and red-tailed hawks that come by pretty regularly. Time passes and 9:00 AM comes and goes, without a visit from the 9:00 AM coyote, who usually shows up like clockwork, making his rounds of the colony while the prairie dogs all stand at the lips of their burrows and call in alarm until he gets close, after which they disappear inside the holes, leaving him to stick his nose in and take it out without a prairie dog in his mouth. Sometimes he is successful at catching an unwary prairie dog, usually by making a rush at a group of animals and having a flustered prairie dog panic and run away from his burrow rather than inside. But today the coyote seems to have gone elsewhere.

Finally, one of us spots something moving in the distant haze at the edge of the field. We all look with our binoculars. It looks like a German shepherd, solidly built with a long fluffy tail, not at all like the slender and skinny coyotes who come hunting prairie dogs. It is walking like a dog, confident, not the furtive run, stop, and look around that we have come to associate with coyotes. We start the video camera and the tape recorder, in anticipation of recording the prairie dogs' responses to a domestic dog.

As the shepherd gets closer, a prairie dog gives an alarm call. A coyote call! We are thrilled. We have been trying to see if the animals would make a mistake in their calls, and now apparently they have mistaken a German shepherd for a coyote. Not that it is hard to do, because coyotes and shepherds look fairly similar, and it would be easy to mistake one for the other. We start mulling over in our minds what this might imply, that there may be limits to prairie dog perception that do not allow them to differentiate between simi-lar-looking predators, and even though domestic dogs have relatively little success in catching prairie dogs, perhaps from the prairie dogs' perspective, it is better to mistake a dog for a coyote than it is to mistake a coyote for a dog. And here at last we have an example of a mistake, where the prairie dogs apparently can't tell the difference between a dog and a coyote.

As one prairie dog gives the coyote call, a chorus of other animals joins in. Prairie dogs that were below ground in their burrows come up and stand on the burrow rims, standing on their hind legs, stretched out and straining to see the predator. Where a few moments ago there were around 20 animals feeding and running around, suddenly there are close to 100 prairie dogs, standing like pegs. The cry is taken up from around the colony. The sound is something like a group of birds chirping, not the loud bark of a dog that one would associate with the name "prairie dog." When visitors to our colony sites are asked if they can hear the alarm calls, a common reply is, no, but I can hear some birds chirping loudly in the background.

The dog comes closer. A moment of doubt and confusion arises among us. Is this really a dog? We look closely through the binoculars. We see that the muzzle is thin, thinner than that of a German shepherd. The tail is more orange, and the legs are more slender. We all come to the same conclusion. This is not a dog, it is a coyote! The prairie dogs were right, and we were wrong. Later, we would spend some time puzzling about how the prairie dogs could tell that it was a coyote at a distance when we could not tell, even with the help of our binoculars.

In retrospect, we could see that natural selection would favor making these minute distinctions. A prairie dog that mistook the coyote for a domestic dog would probably find himself as the main course of a coyote lunch. It is vital for the prairie dogs to be able to make these distinctions, often a literal matter of life or death. For us human observers, it is not vital at all, a matter of passing curiosity, because we know that neither the coyote nor the German shepherd is going to have us on the menu.

5

Population Biology of Prairie Dogs

The study of population biology allows us to understand the ebb and flow of populations of species through time. In cases of critically declining species, like the black-tailed, Utah, Mexican, white-tailed, and Gunnison's prairie dogs, an understanding of population biology becomes a crucial tool in assessing the health of diminishing populations and predicting the threat of extinction. Unfortunately, although a solid understanding of the dynamics of prairie dog populations is sorely needed, we only have a basic understanding of the population biology at present. This chapter summarizes what is known about the population biology of prairie dogs while opening the door for further research on the population dynamics of these critically declining species.

In the first section of this chapter we give an introduction to the life history traits that are important in determining the population biology of all species. We discuss the fundamental parameters that govern population growth like age structures, rate of increase, migration rates, carrying capacity, and sex ratios. We also discuss the importance of studying metapopulations, which are groups of populations in any particular region.

In the second section we identify the actual value of these parameters for prairie dogs. Population biologists have used the above parameters to build population growth models for many populations of organisms including grizzly bears, deer, elk, and wolves. Population models can be used to explore the effects of small populations (the Allee effect), predation, habitat fragmentation, and diseases on the species in question. Creating population growth models for prairie dogs would be one way to explore the effects of plague, poisoning, shooting, and habitat loss on prairie dog populations, but lack of information for some basic life history parameters have precluded the creation of a solid population growth model as of yet.

The third section of this chapter explores the population genetics of prairie dogs. As sedentary, colonial rodents, who are bound to a burrow system for protection from predators, prairie dogs face a significant challenge in maintaining genetic variability. Throughout the evolutionary history of the prairie dog many behavioral strategies have developed to maintain genetic variability. For example, female Gunnison's prairie dogs will often mate with an average of five different males, most of which reside outside the female's home territory (Travis et al. 1996). In addition, female black-tailed prairie dogs will consistently discourage mating attempts from fathers and brothers (Hoogland 1995). Information on levels of inbreeding and genetic relatedness, and the distance and rate of emigration and immigration present a clear picture of the challenges posed to prairie dogs in maintaining genetic variability.

In the past prairie dogs faced only natural challenges to genetic diversity posed by colonial life and low dispersal capabilities. Today, many other factors have combined to further reduce genetic variability in prairie dog populations. Habitat fragmentation has caused an abundance of small isolated colonies in areas where there were historically immense expanses of prairie dogs. Additional challenges to genetic variability include epidemics of sylvatic plague, poisoning campaigns, and recreational shooting. Our last section of this chapter uses the life history characteristics and information on population genetics outlined in the previous sections to attempt to understand the impact that these new disturbances have on the future health of prairie dog populations.

Demography and Populations

Population biology is the study of how populations of organisms change through time. Before we discuss population biology of prairie dogs we need to define some basic guiding principles of the discipline. First and foremost is the definition of a population. A population is simply a group of plants, animals, or other organisms of the same species that live together. In some cases the actual determination of populations is rather challenging, especially in the case of solitary animals like mountain lions and grizzly bears that live in individual territories but interact to reproduce. In this case one must define where the interactions between a group of grizzlies stop and where another interactive group starts and this is often very time consuming and nebulous. Fortunately, in the case of prairie dogs, a population is more easily defined as a colony and the size of the colony population can be tracked through time.

The growth and decline in populations of all species are due to four major factors: births, deaths, emigrations, and immigrations. If births and immigrations exceed deaths and emigrations, the population will grow. If deaths and emigrations exceed births and immigrations, the population will decline. If births and immigrations balance deaths and emigrations, the population will stay static. Thus, because population biology is the study of how populations change, it is essentially the study of births, deaths, emigration, and immigration and the factors that affect these four variables (Wilson and Bossert 1971).

In a textbook world, we can simplify population growth by assuming that there is no emigration and immigration and that birth rates and death rates are only affected by population size. In other words, the more organisms in the population the more births there will be and the more deaths there will be. However, organisms do not live in a textbook world, and immigration and emigration have to be taken into account. This results in the standard equation of population growth: $\triangle N=(B + I) - (D - E)$, where $\triangle N$ is the change in population size, B is the number of births and D is the number of deaths, I is the number of immigrations into the population, and E is the number of emigrations from the population. As the population size (N) becomes larger, we can expect that there will be more births and deaths, simply because there are more individuals who are reproducing and who are dying. We can see that B and D are proportional to the population size N, so that $B = bN$ and $D = dN$, where b is the average birth rate and d is the average death rate per individual per unit time, if we assume that the number of immigrations and emigrations is zero. Converting this to a growth rate we get: $\triangle N/\triangle T = (b - d)N$ The relationship $(b - d)$ is often further simplified into an intrinsic rate of increase called r, making $\triangle N/\triangle T= rN$. This equation for predicting changes in populations over time assumes that only death rates and birth rates affect population growth *and* that nothing affects birth and death rates.

In the real world, there are many environmental factors that can affect population size directly, through changing birth and death rates, or indirectly, through changes in emigration and immigration. Birth and death rates can be affected by amounts of resources available to a population (for example, food, water, or shelter) as well as weather conditions (both severe droughts, severe winters, and extremely favorable conditions). Death rates can be affected by competition, predation, habitat destruction, and epidemics, among other things. Changing levels of emigration and immigration can also affect population growth. Emigration and immigration are affected by distances between populations, the opening or closing of travel routes between populations, the appearance or disappearance of neighboring populations, and other factors such as population density and social structure.

In addition to environmental factors, differences in mortality rates, birth rates, and emigration/immigration rates due to age and sex can contribute to differences in the growth of the population; this is called stage, age, or size-structured growth. For example, juvenile prairie dogs tend to have higher mortality rates than adult prairie dogs and female prairie dogs tend to live longer than male prairie dogs (Hoogland 1995). If the sex ratio of a colony is biased toward non-reproductive juveniles the death rate of a colony may be higher and reproduction lower, or if there are more females than males death rates may be lower but birth rates may be affected by low numbers of males. The most important lesson to learn from age-structured growth is that we cannot treat all prairie dogs the same—there is subtlety and variation between different classes of organisms within the same population and these differences have a big effect on birth, death, and transfer between populations that should not be overlooked when attempting to understand the growth of populations over time.

Carrying Capacity

There are many outside environmental factors that affect population growth. In the case of prairie dogs the four most important factors are plague epidemics, poisoning, predation (both natural and human shooting), and habitat destruction. We will use habitat destruction here to illustrate how changes in the amount of resources available to a population in the surrounding environment affect population growth. Resources for a population are finite. Examples of resources are food, water, shelter, and habitat. As a population grows it contains more individuals that in turn use more resources. As fewer resources become available the number of births in a population usually

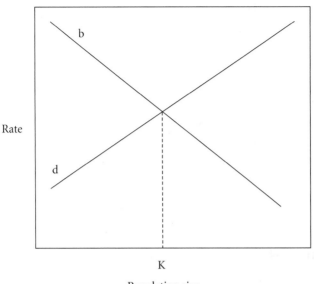

Figure 5.1. Carrying capacity of a species: b = birth rate; d = death rate; K = carrying
capacity, the point where b = d and the population is in equilibrium with its
environment. To the left of K there is a surplus of resources, births exceed
deaths, and the population grows. To the right of K there is a shortage of
resources, deaths exceed births, and the population declines. K is an ideal point;
most natural populations never truly attain K for long periods but rather
fluctuate slightly from one side of K to the other through time.

decreases and the number of deaths eventually must increase, which slows
population growth and eventually decreases population. The point at which
birth rates balance death rates is called the carrying capacity or K. Before K
there are plenty of resources, births exceed deaths and the population grows.
After K there is overcrowding and limited resources, deaths exceed births and
the population decreases. At K birth rates and death rates are equal and the
population is at equilibrium with its resources (Figure 5.1), assuming that
there are no deaths due to predation or disease.

Changes in resource levels over time can push a population above or below
carrying capacity. For example, in the case of prairie dogs, destruction of
prairie habitat decreases a fundamental resource that provides food and shelter to prairie dogs. This could push prairie dog populations over their carrying capacity and can lead to a decline in prairie dog populations. Conversely,

restoring large expanses of prairie could lead to an increase in prairie dog populations by letting them fall below carrying capacity.

Metapopulation Biology

There are many levels of complexity to life that contribute to the persistence of a species through time and, if taken far enough, the persistence of life itself (Figure 5.2). The continued health and well-being of individuals within a population is a basic essential to population persistence. Obviously, if there are no individuals, a population cannot exist. Individuals are also the basic carriers of the genetic code within a population: each individual contributes to the genetic heritage of the species, and selection only operates on the individual. Hence, every life is precious. However, while the demise of an individual may affect the genetic heritage of a species to some extent, it will most likely not affect the survival of an entire population or species, unless that population or species has only a few members.

After the individual, populations are a larger unit of complexity within a species. Populations are made up of individuals of the same species. The interactions of the population members with each other and with their local environment determine the persistence of populations through time.

The demise of a population carries more risk of extinction to a species than the death of an individual, but this risk depends on how much the population contributes to the well-being of the species as a whole and the unique genetic components of the individuals who make up the population. The relative weight of contribution to a species for each population can only be uncovered by looking at the next level of complexity, the metapopulation.

A metapopulation is best defined as a population of populations, in other words, a group of several, interacting local populations or subpopulations that are linked together by immigration and emigration (Gottelli 1998). So the study of metapopulations is the study of interactions between populations of the same species. As we discussed earlier, emigration and immigration are two important variables in determining population growth within a single population. Emigration and immigration are only fully understood by acknowledging neighboring populations. It is only at this level of complexity that the extinction risk of a species can be adequately tackled. Metapopulation biology breaks the assumption of a closed population that was present in the

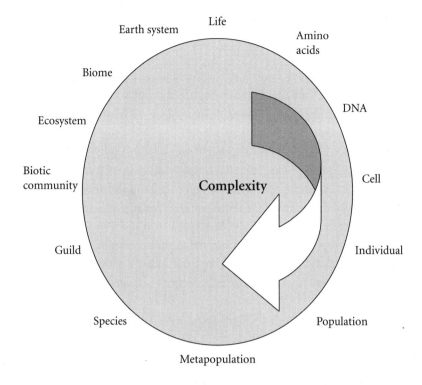

Figure 5.2. Life's different levels of complexity. All levels contribute to the persistence of a species through time, and the interaction between species and between species and the physical environment ultimately contributes to the continued existence of life itself. Complexity increases from amino acids to the earth system (interactions between all biomes and physical systems). Amino acids form proteins, which are the building blocks of life. DNA is the basic carrier of the genetic code essential to life. Individuals reproduce asexually or sexually, passing on the DNA to future generations. Populations are groups of individuals that contribute to diversity (sexual reproduction) and provide safety for individuals to persist. Metapopulations further contribute to diversity and strengthen populations through emigration and immigration potential. Guilds are a group of different species that make their living in the same way—for example, seed-feeding birds. Biotic communities are communities of plants and animals including trophic connections between them but excluding physical factors. Ecosystems are biotic communities that include interactions of the biotic community with the physical environment (i.e., climate, substrates). Biomes are major ecosystem types. This figure is cyclical because interactions in the earth system renew the basic elements of life and allow life to continue to persist.

population models described previously and deals with links between populations, and ultimately with extinction risk. Common terms in metapopulation
equations are local population extinction verses local population persistence;
regional population extinction verses regional population persistence; rescue
effect, sources, and sinks; and probability of local and regional extinction. All
these are crucial terms to a conservation biologist seeking the best conservation actions to prevent extinction of a species.

In terms of prairie dogs, or species that are not very mobile and that carry
a high risk of extinction due to habitat fragmentation or high rates of mortality, understanding metapopulation biology is paramount in understanding
extinction risk and appropriate conservation action. Connectivity between
prairie dog colonies significantly influences the ebb and flow of single populations, of metapopulations, and ultimately the species as a whole. This
influence carries both positive and negative potential through providing
gene flow, disease pathways, resource flexibility, and pathways to population
expansion.

In the following section we will discuss demographics of prairie dog populations. All of the demographic factors contribute to population growth of
prairie dogs by affecting the four fundamental factors: birth, death, emigration, and immigration.

Prairie Dog Demographics

In an ideal world, we would be presenting a population model for prairie
dogs that could be used to predict the effects of habitat destruction, plague
epidemics, recreational shooting, and predation on prairie dog populations.
This model would integrate population and metapopulation variables and
could be used to predict to what extent prairie dog populations are in danger
of becoming extinct. This would provide an objective evaluation of which
prairie dog species need to be placed on the Endangered Species list and
have recovery efforts implemented. However, this is not an ideal world. The
generation of a population model for prairie dogs, although not impossible,
is very challenging. Unlike larger animals such as elk, bears, and cougars, all
of which have had population models generated for them, prairie dogs are
smaller organisms with shorter life spans. This poses a few major challenges.
Short life spans mean population rates that change more quickly in the face

of environmental change. Survival rates and litter sizes of ground squirrels change with the length of growing season, elevation of the town, poorly understood disease dynamics, human political machinations, and sometimes inexplicably between years (Bronson 1979; Murie 1985; Smith and Johnson 1985). Prairie dog populations can decrease rapidly, but they increase relatively slowly and somewhat unpredictably to habitat changes, sometimes within the same season (Morton and Sherman 1978; Dobson and Kjelgaard 1985; Dobson et al. 1997, 1998; Smith and Johnson 1985). Life history traits of white-tailed prairie dogs are highly variable between colonies, and within the same colony between years. There is no clear answer for what contributes to these changes. In addition, age structured differences are complicated and vary between colony and between species. Juvenile body mass of white-tailed prairie dogs, but not adult prairie dogs, is affected by temporal and spatial habitat variation (Bakko and Brown 1967). Ground squirrels are also dynamic reproducers, varying reproduction temporally based on resource availability. Because of this extreme variation, some authors have asserted that life tables and population models are not appropriate for prairie dogs (Bakko and Brown 1967).

Separating demographic variables by species can eliminate some variation in demographic characteristics. There are varying amounts of information on life characteristics depending on the species of concern. Black-tailed prairie dogs are the most studied and there is a complete repertoire of information on breeding, age structures, survival rates, and some dispersal information available in the existing literature (Table 5.1). Gunnison's prairie dogs and white-tailed prairie dogs have less information than black-tailed prairie dogs, but there is still quite an extensive body of literature describing the demographics of these two species. Utah prairie dogs and Mexican prairie dogs have a paucity of information available. Ironically, these are the two species listed as threatened and endangered respectively, with the lowest population sizes of all species in the genus, a telling confirmation of how ill-prepared we are in terms of population biology knowledge to protect this declining genus.

In all species of prairie dogs, estrus of females is very short, averaging only one day per female. Females may stagger their times of estrus to avoid competition with more dominant females but, in all, the reproductive period for prairie dogs is very short. Mexican prairie dogs are the only species that have

Table 5.1. Demographic characteristics of prairie dogs by species.

Species	Parameter	Value	Age/Sex	Notes	Reference
Cynomys ludovicianus	Hibernation season	No hibernation			
Cynomys mexicanus	Hibernation season	No hibernation			
Cynomys leucurus	Hibernation season	Aug./Sept. to late Feb./ early–mid-March	Male Female	Males emerge two–three weeks before females	Menkens and Anderson 1989
Cynomys gunnisoni	Hibernation season	Oct. to early March/ late April			Slobodchikoff and Perla (pers. obs.)
Cynomys parvidens	Hibernation season	Winter			USFWS 1991
Cynomys ludovicianus	Life span	4 years 6 years	Male Female		Hoogland et al. 1987
Cynomys mexicanus	Life span	Unknown			
Cynomys leucurus	Life span	Unknown			
Cynomys gunnisoni	Life span	3 years 4 years	Male Female		
Cynomys parvidens	Life span	5 years 8 years	Male Female		USFWS 1991
Cynomys ludovicianus	Survival rate	.50 .50	1 year old males 1 year old females	Infanticide is primary cause of infant mortality (51% of litters)	Hoogland et al. 1987
Cynomys mexicanus	Survival rate	Unknown			

Species	Parameter	Value	Sex/age class	Notes	Reference
Cynomys leucurus	Survival rate	.09–70 .09–40 .20–63 .08–50	Adult male Juvenile male Adult female Juvenile female	Averages, highly variable	Menkens and Anderson 1989
Cynomys gunnisoni	Survival rate	35%–80% 36%–66%	Juvenile male Juvenile female		Rayor 1985
Cynomys parvidens	Survival rate	Unknown			USFWS 1991
Cynomys ludovicianus	Age to first reproduction	2	Male, female	3% males, 9% of females reproduce after one year	Hoogland et al. 1987
Cynomys mexicanus	Age to first reproduction	Unknown			Backo and Brown 1967
Cynomys leucurus	Age to first reproduction	1 year	Male, female		Menkens and Anderson 1989
Cynomys gunnisoni	Age to first reproduction	11 months	Female	Males more variable	Hoogland 1997
Cynomys parvidens	Age to first reproduction	1 year			
Cynomys ludovicianus	Dispersal age	1–5 years	Male, female	Males are primarily yearlings, while females can be older	Knowles 1985 Garrett and Franklin 1988
Cynomys mexicanus	Dispersal age	Unknown			
Cynomys leucurus	Dispersal age	1 year	Juvenile male	Juvenile males are dominant dispersing class	Menkens and Anderson 1989
Cynomys gunnisoni	Dispersal age	Over 1 year	Male, female		Rayor 1985
Cynomys parvidens	Dispersal age	Unknown			USFWS 1991

Table 5.1. (continued)

Species	Parameter	Value	Age/Sex	Notes	Reference
Cynomys ludovicianus	Sex ratio	42%–59% male	Pups		Hoogland et al. 1987
Cynomys mexicanus	Sex ratio	Unknown			Menkens and Anderson 1989
Cynomys leucurus	Sex ratio	1:1	Juvenile	Female:Male	Menkens and Anderson 1989
		3:2	Adult		
Cynomys gunnisoni	Sex ratio	1:2	Pups	Female:Male	Fitzgerald and Leichleitner 1968
		1:6	Adults	Female:Male	
Cynomys parvidens	Sex ratio	Unknown			USFWS 1991
Cynomys ludovicianus	Litter size	4.4			Knowles 1987
		3.11 ± .08			Foltz et al. 1988
Cynomys mexicanus	Litter size	4			Pizzimenti and McClenaghan 1974
Cynomys leucurus	Litter size	5.48	Adult	Average	Stockard 1929
Cynomys gunnisoni	Litter size	3.71 ± 1.18			Hoogland 1997
Cynomys parvidens	Litter size	1–6			USFWS 1991
Cynomys ludovicianus	Avg. weight	150.9 ± 1.5	Pup male	Grams	Foltz et al. 1988
		145.8 ± 1.6	Pup female		Hoogland et al. 1987
		600–1,200	Adults male and female		
Cynomys mexicanus	Avg. weight	1,200	Adult male	Grams	Pizzimenti and McClenaghan 1974
		900	Adult female		
Cynomys leucurus	Avg. weight	682 ± 80.2 g	Male spring	Adults average (g)	Backo and Brown 1967
		1,400–1,600	Male fall		

Species	Measurement	Category	Value	Unit	Reference
Cynomys gunnisoni	Avg. weight	Female spring	694.2 ± 111.7	Averages (g)	Fitzgerald and Lechleitner 1968
		Female fall	1,000		Rayor 1985
		Male juvenile fall	960.8 ± 72.1		
		Female juvenile fall	695 ± 52.9		
Cynomys parvidens	Avg. weight	Adult male	816		USFWS 1991
		Adult female	644		
		Pup male	225.9		
		Pup female	221.1		
		Adult	200–900	Grams	
Cynomys ludovicianus	Breeding season		Early March–early April		Knowles 1987
Cynomys mexicanus	Breeding season		Early Jan.–June		Pizzimenti and McClenaghan 1974
Cynomys leucurus	Breeding season		Late March, early April–end of April		Stockard 1929
Cynomys gunnisoni	Breeding season		Mid-March–early April / Late April–early May		Hoogland 1997 / Fitzgerald and Leichleitner 1968
Cynomys parvidens	Breeding season		Unknown		USFWS 1991
Cynomys ludovicianus	Gestation period		34.8 days ± .1		Foltz et al. 1988 / Hoogland et al. 1987
Cynomys mexicanus	Gestation period		30 days		Pizzimenti and McClenaghan 1974
Cynomys leucurus	Gestation period		27–33 days		Stockard 1929
Cynomys gunnisoni	Gestation period		29.3 days ± .53 days		Hoogland 1997

Table 5.1. (continued)

Species	Parameter	Value	Age/Sex	Notes	Reference
Cynomys parvidens	Gestation period	30 days			Knowles 1985
Cynomys ludovicianus	Dispersal season	Late May, early June to early August (10-week window)	Adult males and females 1–5 years old to other towns	Within-town dispersers were only 1-year-old males	Garrett 1982
Cynomys mexicanus	Dispersal season	Unknown			
Cynomys leucurus	Dispersal season	Early April	1-year-old males		Bakko and Brown 1967
Cynomys gunnisoni	Dispersal season	May–June			Rayor 1985
Cynomys parvidens	Dispersal season	Unknown			USFWS 1991
Cynomys ludovicianus	Immigration ratio	6%–42% of population	% of town population from immigration population		Knowles 1985
Cynomys mexicanus	Immigration ratio	3.3 prairie dogs/colony/generation			McCullough and Chesser 1987
Cynomys leucurus	Immigration ratio	Unknown			
Cynomys gunnisoni	Immigration ratio	Unknown			
Cynomys parvidens	Immigration ratio	Unknown			USFWS 1991
Cynomys ludovicianus	Litters/lifetime	4	Female	Derived from info on age to reproduction and lifespan; 63%–74% of females breed/yr.	Foltz et al. 1988 Hoogland et al. 1987 Knowles 1987
Cynomys mexicanus	Litters/lifetime	Unknown			
Cynomys leucurus	Litters/lifetime	Unknown			

Species	Measurement	Value	Notes	Reference
Cynomys gunnisoni	Litters/lifetime	3	Derived from info on age to reproduction and lifespan	USFWS 1991
Cynomys parvidens	Litters/lifetime	Unknown		
Cynomys ludovicianus	Dispersal distance	0–10 km; average 2.4 km	Dispersers primarily used roads (97%), distances over 3 km 77%, male dispersers	Knowles 1985 Garrett 1982 Roach et al. 2001
Cynomys mexicanus	Dispersal distance	Unknown		
Cynomys leucurus	Dispersal distance	Unknown		
Cynomys gunnisoni	Dispersal distance	Less than 13 km		Fitzgerald and Leichleitner 1968
Cynomys parvidens	Dispersal distance	Unknown		USFWS 1991
Cynomys ludovicianus	Length of lactation	43.4 ± .8 days		Foltz et al. 1988
Cynomys mexicanus	Length of lactation	40 days		Pizzimenti and McClenaghan 1974
Cynomys leucurus	Length of lactation	7 weeks	Female	Bakko and Brown 1967
Cynomys gunnisoni	Length of lactation	38.6 ± 2.08	Female	Hoogland 1997
Cynomys parvidens	Length of lactation	35 days		USFWS 1991
Cynomys ludovicianus	Length of estrus	1 day (5–6 hours)	Female	Hoogland 1982
Cynomys mexicanus	Length of estrus	Unknown		
Cynomys leucurus	Length of estrus	Unknown		
Cynomys gunnisoni	Length of estrus	4–10 hours		Hoogland 1997
Cynomys parvidens	Length of estrus	Unknown		USFWS 1991

been reported as reproducing for a longer period of time (Pizzimenti and McClenaghan 1974), although further study may show that their reproductive period is as limited as the other four species. Short reproduction windows, coupled with relatively low litter sizes and longer periods to sexual maturity, make prairie dogs an exception in the rodent world and lead to comparatively lower population growth rates than other rodent species.

Dispersal is often defined as a permanent movement away from the home range or place where an animal was born (Lidicker and Stenseth 1992). Dispersal characteristics tend to vary highly with prairie dog species, but dispersal time usually coincides with breeding season. Gunnison's juvenile female and male prairie dogs are the dominant dispersers, white-tailed juvenile males are the primary dispersers and black-tailed prairie dog adults, both female and male, disperse. Dispersal distances are short, averaging 2–3 kilometers (Garrett 1982), and males tend to disperse longer distances than females. We have compiled the information available on life characteristics, along with references, for each species in Table 5.1.

For all species, the least studied population characteristics are metapopulation characteristics, most importantly dispersal, emigration, and immigration. This reflects the bias of most prairie dog researchers to single populations and the difficulty of studying dispersal. Conservation of prairie dogs is a new idea, and thus metapopulation research has not reached its potential for this genus. Genetic research (dealt with in subsequent paragraphs) is slowly breaking into a metapopulation view, but more work is sorely needed. Further research into this area is imperative for conservation efforts.

Genetics of Prairie Dogs

The prairie dog, a sedentary, communal animal, faces great challenges in maintaining genetic diversity and minimizing inbreeding effects (inbreeding depression). In addition, low dispersal capabilities (2–3 kilometers average dispersal distance) often produce high levels of heterogeneity between populations (Chesser 1983). Maintaining a balance between successful levels of genetic diversity and the benefits, in terms of safety, of living in a tight social structure is the single most important challenge prairie dogs face in terms of population biology.

There are many genetic measures of diversity that are used by population geneticists, including coefficient of relatedness, genetic distance, and genetic differentiation. The coefficient of relatedness (r) measures levels of relatedness between two individuals by comparing how much genetic material these two individuals have in common (Hamilton 1964). A brother and sister have a coefficient of relatedness of 0.5, sharing half their genes. A clone would have a coefficient of relatedness of 1.0, sharing all its genetic material with its sister individual. Genetic distance and genetic differentiation measures are most commonly measured using F-statistics and allele frequency data. These measurements are the converse of coefficient of relatedness values measuring the degree of separation rather than the degree of similarity.

In general, prairie dogs seem to have high levels of genetic relatedness within a colony. Gunnison's prairie dogs are apparently the most related, with a coefficient of relatedness ranging from 0.55–0.61, suggesting that any two individuals randomly chosen within a colony are going to be more closely related genetically than siblings normally are related in a population that is not inbred (Travis et al. 1997). Mexican prairie dogs are slightly less related within a colony than Gunnison's prairie dogs, with relatedness levels averaging 0.47 (McCullough and Chesser 1987). In Gunnison's prairie dogs, the high level of relatedness might come from "boom and bust" cycles caused by plague. Plague can kill up to 99% of all the animals in a colony, the "bust" part of the cycle, leaving a few animals to start up the colony again. This leads to the Founder Effect, which is a situation that arises when a population starts growing from one or a few animals—the gene alleles that these "founder" animals have are the ones that are going to be represented in the new population. Other, different genetic alleles that might have been present in the animals that died, and any genetic diversity that might have existed in the original population, are lost. The Founder Effect can produce a bottleneck in genetic diversity. The new population then starts to increase in numbers (the "boom" part of the cycle), but with a much more limited set of genetic alleles. Although such "boom and bust" cycles can potentially come from plague in Gunnison's prairie dogs, the Mexican prairie dogs do not currently have plague, and the high levels of genetic similarity might come from limitations on gene flow produced by the social system (see chapter 3), or by poisoning, or the long-term persistence of a small population, which also lead to a gradual erosion of genetic diversity.

The levels of relatedness within black-tailed prairie dog colonies are not clear (Dobson et al. 1997; Dobson et al. 1998). One study on a number of black-tailed colonies in New Mexico found high levels of within-colony related-ness. Other studies of a colony at Wind Cave, South Dakota, suggested lower levels of within-colony relatedness. However, it appears that black-tails have higher levels of relatedness within the social groups, the coteries, than within the colony as a whole. Despite the higher levels of relatedness there has been no evidence of inbreeding depression detected (Hoogland 1992). A study of the genetic structure of 13 colonies of black-tailed prairie dogs in the Central Plains Experimental Range and Pawnee National Grassland in eastern Colorado showed little evidence of inbreeding within the colonies (Roach et al. 2001).

Measures of heterogeneity reflect how much gene flow occurs between populations. Populations that are completely isolated and have no gene flow from neighboring populations have high values of genetic differentiation approaching 100% (or, the scale can be measured from 0 to 1, where 0 means complete gene flow and no genetic differentiation between populations, and 1 means no gene flow and complete genetic differentiation). Gene flow between populations is facilitated by dispersing animals—the more dispers-ers there are, the higher the gene flow between populations.

Table 5.2.   Coefficients of relatedness and other genetic information on prairie dogs by species.

Species	Genetic variable	Level: Value	Reference
Cynomys ludovicianus	Differentiation	b/w coteries: 0.166	Dobson et al. 1998
		b/w coteries: 0.227	Chesser 1983
		b/w colonies: 0.103	Chesser 1983
		b/w colonies: 0.115	Daley 1992
		b/w colonies: 0.118	Roach et al. 2001
		b/w colonies: 0.194	Trudeau et al. 2004
	Relatedness (F_{IS})	w/in colony: 0.017	Roach et al. 2001
Cynomys mexicanus	Differentiation	b/w colonies: 0.07	McCullough and
	Relatedness (F_{IS})	w/in colony: 0.47	Chesser 1987
Cynomys leucurus		Unknown	
Cynomys gunnisoni	Relatedness	w/in colony: 0.55–0.61	Travis et al. 1997
	Differentiation	b/w colonies: 0.11	
Cynomys parvidens		Unknown	

The three prairie dog species that have been studied in terms of their genetics show some genetic differentiation between colonies. Genetic differentiation for Gunnison's prairie dogs averages 11% sampled over a scale of less than 20 kilometers (Travis et al. 1997). Mexican and black-tailed prairie dogs average 7% and 4.9%–19.4% respectively, sampled over broad geographic regions (Chesser 1983; McCullough and Chesser 1987; Daley 1992; Dobson et al. 1998; Roach et al. 2001; Trudeau et al. 2004). A breakdown of genetic diversity measures for different prairie dog species is shown in Table 5.2. Prairie dogs have a low number of dispersers between colonies. Genetic studies suggest that the observed heterogeneity in prairie dog colonies could be produced by as few as one to five dispersers per generation from other colonies in black-tailed prairie dogs and two dispersers per generation in Gunnison's colonies (Garrett 1982; Chesser 1983; Travis et al. 1997). This low number of dispersers is probably the result of the difficulty of new animals integrating themselves into the existing social groups within a colony.

Population Biology and Conservation of Prairie Dogs

A good understanding of prairie dog population biology on the individual, population, metapopulation, and species level is essential to making wise decisions about conservation of this genus. Unfortunately, the research bias concerning population biology has not traditionally been concerned with prairie dog conservation and therefore lacks strength in terms of knowledge of significantly rare and threatened species (i.e., the Utah and Mexican prairie dogs) as well as metapopulation dynamics (specifically mechanisms, rates, and limitations of emigration and immigration). However, genetic research provides a strong tool for discovering relationships between colonies of prairie dogs and for predicting gene flow rates. Such research is increasingly being turned toward conservation themes.

More information on emigration and immigration between populations of prairie dogs will increase our understanding of the effects of habitat fragmentation, disease transmission (most importantly plague), recreational shooting, and poisoning deaths. Prairie dog populations are extremely fragmented compared to historical ranges, and prairie dog town sizes are smaller. Houses, parking lots, and agricultural fields not only decrease existing prairie dog habitat, but may also serve to block movement corridors between colonies. Blocking movement corridors could tip the delicate balance that prairie dogs have successfully maintained between living in social groups and avoiding the effects of inbreeding depression. An understanding of movement corridor lengths, types, and importance to population diversity is sorely needed.

On the other hand, plague epidemics are extremely devastating to prairie dog populations, and the increased habitat fragmentation and blocked movement corridors between prairie dog towns could provide some level of protection against the spread of this destructive disease. However, it is equally likely that fragmentation could be increasing the impact of the disease through limiting re-colonization of prairie dog towns hit by plague. As plague can travel from town to town through predator vectors (e.g., on coyotes that carry infected fleas), disruption of movement corridors that prairie dogs use in dispersing may not prevent spread of the disease as predator vectors may use different and longer movement corridors. Yet re-colonization could be blocked by disruption of prairie dog movement corridors, leading to a faster decline in metapopulations. More research on plague transmission, its relationship with prairie dog habitat fragmentation, and its long-term effect on metapopulations would be a fascinating and extremely important topic to explore.

With the advent of new genetic techniques, and with determined efforts to accumulate data concerning population parameters, informative models of prairie dog population dynamics may be possible in the near future. In any case, exploring the diversity of population parameters between colonies and between species of prairie dogs, and uncovering the factors (environmental, human-caused, or other) that contribute to this variation, is a necessary exercise that will contribute greatly to our understanding of what prairie dog populations require for their persistence in future generations.

6

The Ecology of Prairie Dogs

In recent studies of grassland ecology, the prairie dog has arisen again and again as a key player in the grassland ecosystem. Most ecologists believe prairie dogs are keystone species in North American grasslands—like the keystone of an arch, prairie dogs hold the structure of the prairie ecosystem together. In a more technical definition of keystone species, prairie dogs and their actions must have a more integral role in maintaining the health of this system than one would expect by their population size (Miller et al. 1994), and the ecological function of prairie dogs must be largely unduplicated by other species (Kotliar 2000). There is disagreement as to whether we know enough to call prairie dogs keystone species in the true sense of the term (Stapp 1998). There is widespread consensus that the effects of prairie dogs in grasslands are far-reaching and pervasive in all levels of the ecosystem (Kotliar et al. 1999).

In this chapter, we focus on the ecology of the prairie dog in North American grasslands. We start by defining grasslands. Then we ask: What are some of the roles that prairie dogs play in this system? How do prairie dogs contribute to the continuing function of grasslands? What are the interconnections between prairie dogs and other species of the North American prairie? We end the chapter by discussing the strengths of these interconnections and how vital they are to the healthy functioning of grasslands in North America.

The Great Plains Prairie

We cannot discuss prairie dog ecology without mentioning the Great Plains prairie. As its name "prairie dog" suggests, the life of a prairie dog is lived completely within North American grasslands, of which the Great Plains prairie system makes up the majority. The range of prairie dog habitat is centered on the Great Plains with extensions into the western outlying grasslands of this system in Arizona, Utah, and New Mexico, and southward extensions into northern and central Mexico.

The historical definition for the Great Plains or Great Basin grasslands is an open landscape in which grasses are dominant and provide a continuous, largely uninterrupted cover (Brown 1982). In the context of geologic time the Great Plains prairie is fairly young. Prior to the Pleistocene, a woodland-grassland mosaic covered what is now the Great Plains (Axelrod 1985). It was not until well into the Holocene epoch, after most of the Pleistocene glacial ice had retreated, that the environment could support the large contiguous grasslands that formed the North American prairie (Steinauer and Collins 1996).

From the end of the Pleistocene until recent times the Great Plains prairie formed a relatively continuous span of grassland stretching from southern Canada to northern Mexico and from the plains states in the east to the eastern edges of Arizona and Utah in the west (Mulhern and Knowles 1995). Now it is very fragmented. However, it still retains vestiges of the three major prairie subdivisions that were widespread in the past: short-grass, mixed-grass, and tall-grass prairie.

The short-grass prairie is the westernmost extension of the Great Plains system. It runs from the eastern side of the Rocky Mountains eastward for about 320 kilometers (190 miles) until it merges with the mid-grass prairie and forms a long north-south band stretching from northern Mexico to central Alberta in Canada (Weaver et al. 1996). Outlying patches of short-grass prairie also exist in Arizona (Brown et al. 1980).

The short-grass prairie is the most arid of all prairie subtypes with annual precipitation hovering near 375 millimeters (15 inches) in some areas (Brown et al. 1980), and lower in many other areas. The low amount of precipitation is one reason for the short grasses that characterize this prairie type. Dominant species vary along the north-south gradient, but include blue grama, buffalograss, hairy grama, weedy grasses like sand dropseed and squirreltail, as well as small forbs. The short-grass prairie is very tolerant of grazing pressure (Milchunas 2006; Milchunas et al. 1988), but grazing has altered the native floral composition in much of the system (Brown et al. 1980, Weaver et al. 1996).

The mixed-grass prairie is the central zone of prairie, existing between the short-grass prairie to the west and the tall-grass prairie to the east (Kuchler 1985). This prairie was the central range of the historic bison herds. Precipitation rates are variable along both an east-west and north-south gradient ranging from 300 millimeters (12 inches) in the driest southwestern corner to 800 millimeters (32 inches) in the wettest northeastern corner. The mixed prairie is forb-dominated in terms of diversity. It is dominated in terms of productivity by a handful of perennial grass species. The mixed prairie is more diverse than the other two prairie subtypes and is further divided into a northern mixed prairie covering the Dakotas and Canada, dominated by cool season grasses; a Sandhills complex encompassing areas in Nebraska, defined by sandy substrates; and a southern mixed prairie that covers an area from Kansas to Texas, characterized by warm season grasses and loam or clay soils (Brown et al. 1982; Bragg and Steuter 1996).

The tall-grass prairie is the easternmost prairie subtype and is bounded by deciduous forest to the east. The tall-grass prairie is characterized by higher annual precipitation rates (600–1,000 millimeters (24–40 inches)) and cooler temperatures than the other two subtypes. Tall grasses like big bluestem and Indian grass dominate and shorter grasses are subdominant. The tall-grass prairie has experienced the most declines in range of all subtypes, at or near 100% in many areas (Kuchler 1985; Samson and Knopf 1994; Steinauer and Collins 1996).

Of the three prairie types, prairie dogs tend to occupy short and mid-grass prairies more than tall grass prairies. Gunnison's, white-tailed, and Utah prairie dogs occur mostly in the western prairie states and tend to predominantly inhabit short grass prairies. Mexican prairie dogs are found more often in the southern parts of the Great Plains prairie, inhabiting mid-grass prairies more than the other species. Black-tailed prairie dogs are the most abundant and widespread, living in mid- and short-grass prairies and occasionally in tall-grass prairies.

Ecological Processes in Grasslands

The most influential abiotic processes (processes that are not due to the action of living things) in most grasslands are fire and precipitation. Precipitation is usually lower annually than that of forest and woodland formations and characterized by fairly frequent extreme maximum or minimum precipitation events. Frequent fire cycles and low precipitation rates are dominant abiotic factors in preventing the growth of shrubs and woody plants. In undisturbed grasslands fire cycles often lead to a community that is dominated by a mixture of forbs and grass, rather than just grass (Weaver and Albertson 1956; Humphrey 1958; Wallis 1982; Williams 1982). However, on grasslands heavily grazed by livestock there is less fuel available for natural fire regimes to take place and grasslands often shift to a community dominated by shrubby plants rather than grasses and forbs (Brown 1982). Along with the aforementioned factors, soil type seems to be important in regulating plants and hence, perhaps animal distributions (Weaver and Albertson 1956; Lowe 1985).

While abiotic processes determine the initial conditions for what grows in a plant community, biotic processes (processes that result from the activity of living things) on grasslands are extremely important in the ongoing maintenance and make-up of grassland communities (Lewis 1982). Soil quality and nutrient cycling are regulated by biological processes such as herbivory and burrowing of grassland animals. Studies also show that while abiotic factors such as low precipitation rates may induce formation of grasslands, biotic factors such as grazing pressure and nutrient cycling by burrowing rodents seem to regulate the continuance of grassland habitats (Detling 1988; Whicker and Detling 1989; Munn 1989; Bock and Bock 1993; Weltzin et al. 1997).

As one biotic force, the influence of the prairie dog in maintaining grassland systems is extensive (Tables 6.1, 6.2). Prairie dogs modify the grassland ecosystem through burrowing and grazing and clipping of tall vegetation. They are also prey to a variety of grassland animals including coyotes, raptors, snakes, badgers, black-footed ferrets, and foxes. Burrows act as refuges for burrowing owls, snakes, cottontails, ferrets, invertebrates, lizards, toads, salamanders, and small rodents. Prairie dog activities also change vegetation composition and structure and increase nutrient levels in foliage that attract ungulates like bison and pronghorn. In the following sections we will discuss these influences of the prairie dog on the grassland system in more detail.

Prairie Dogs, Bison, and Pronghorn Antelope

Because prairie dogs are burrowing rodents and herbivores, they exhibit a strong influence that initiates a cascade of effects felt throughout higher

Table 6.1. Species that have clear positive associations with prairie dogs.

Species	Benefit/resource	Type of association
Forb species	Increased nutrients, decreased competition	Population and species diversity increases
Short grasses	Increased nutrients, decreased competition	Population increases but time dependent; decreases in older prairie dog colonies
Native vegetation	Increased nutrients, decreased competition	Increased population and species richness unless colony is grazed
Pogonomyrmex rugosus	Forage (?)	Population increases
Coleopterans	Food	Population increases
Tenebreonids	Habitat	Increased abundance, increased unique species, lower species diversity
Root nematodes	Increased forage	Population increases
Pollinating insects	Forage	High-use of colony
General ground-dwelling arthropods	Habitat, forage	Population increases
Burrowing owl	Nesting, rodent prey	Population and density increases
Mountain plover	Nesting, insect prey	Population and density increases
Ferruginous hawk	Predator	Population increases
Red-tailed hawk	Predator	Population increases
Bald eagle	Predator	Population increases
Golden eagle	Predator	Population increases
Swift fox	Predator, habitat	Semi-obligate
Insectivorous birds	Insect prey	High-use of colony
Exotic birds	Insect prey	High-use of colony
Insectivorous rodents (i.e., *Peromyscus spp., Onchymous spp., Microtus ochragaster, Perognathus spp., Mus musculus, Spermophilus tridecimlineatus*)	Insect prey	Population increases
Black-footed ferret	Predator, habitat	Obligate
Bison	Increased nutrients in forage	High-use of colony
Pronghorn	Increased nutrients in forage	High-use of colony
Herptiles	Burrows for hibernacula and refugia	Population increases
Predatory herptiles	Predator	Population increases

Table 6.2. Prairie dog influences on biodiversity and ecological processes.

Species/process	Effect/level of effect	+/−	Prairie dog activity	Notes
Soil	Aeration/direct	+	Burrowing	
Soil	Nutrient cycling/direct	+	Burrowing/grazing	
Surface soil	Nutrient load/direct (soild mixing)	+/−	Burrowing	If topsoil is rich and underlying soil saline or calcareous, negative effect. If topsoil is poor and underlying soils rich, positive effect.
Plant foliage	Nutrient load/direct and secondary	+	Burrowing/grazing	
Plant community structure	Grazing resistant forms and species/direct	+	Grazing	
	Perennial grasses/direct	−	Grazing	
	Short grasses/direct—often negative, depending on the level of grazing pressure	+/−	Grazing	Time dependent
	Forbs/direct	+	Grazing	
	Vegetation patchiness/secondary	+	Granivory	Prairie dogs eat seeds where they are most abundant (under existing plants), results in plants germinating only at distances from others.
	Overall plant diversity/direct (can be positive depending on grazing pressure)	+/−	Clipping, grazing	Black-tailed prairie dog only
Trees	Encroachment/secondary	−	Granivory, grazing	Prairie dogs prevent forest and shrubland encroachment through eating seeds and young shoots of trees
Foliage dwelling and flying arthropods	Abundance/secondary	−	Secondary effects	Changes in plant community structure on colonies bring more insectivorous predators
Root nematodes	Abundance/secondary	+	Grazing	Grazing increases root mass of plants
Pogonomyrmex rugosus	Abundance/secondary	+	Grazing	Plant community structure changes to accommodate forage needs
P. barbatus, occidentalis	Abundance/direct or secondary	−	Grazing/competition-granivory	Possibly changes in plant community and direct competition with prairie dog diet

Tenebreonid	Richness/secondary	−	Unknown	Perhaps changes in plant community or habitat diversity
Tenebreonid	Abundance/direct	+	Burrowing	
Coleopterans	Abundance/direct	+	Burrowing, grazing	Increases flowering dicot forage
Pollinating insects (?) (especially bees)	Abundance/secondary	+	Burrowing, grazing	
Insectivorous birds	Abundance/secondary	+	Maintenance of openness—easier to locate prey	
Predatory insects	Abundance/secondary	+	Unknown	
Raptors	Abundance/direct	+	Prey	Both prairie dogs and other small mammals
Exotic birds	Abundance/unknown		Unknown	
Burrowing owl	Abundance/direct-secondary	+	Burrowing, grazing	Burrow habitat and increased small rodent prey
Mountain plover	Abundance/direct-secondary	+	Burrowing, grazing, clipping	Nesting and foraging habitat, prey
Insectivorous rodents	Abundance/secondary	+	Burrowing, grazing	Changes in plant community structure attract invertebrate prey
Small mammals	Abundance/secondary	+	Burrowing, grazing	Can be decreased diversity, but increased abundance; but can also increase diversity
Ungulate	Richness/direct	−	Competition	More nutrients, but less biomass (probably a mosaic)
	Grazing/secondary	+	Secondary effect of burrowing, grazing	
Terrestrial predators (coyote, fox, badger, rattlesnake, black-footed ferret, etc.)	Abundance/direct	+	Prey	
Reptiles	Abundance/secondary	+/−	Burrows, grazing, clipping	Provide refugia, hibernacula, and less cover for invertebrate prey

trophic levels. Like giant earthworms, prairie dogs move great quantities of soil below ground and bring deeper soils to the surface. This aerates the soil, increases nutrient cycling and changes soil texture (i.e., the proportion of sand, silt, and clay). As herbivores, prairie dogs affect plant structure and species composition on their colonies by selectively eating certain plants, and causing structural changes in some species through their grazing activities (Coppock et al. 1983a). A combination of grazing and soil cycling effects can increase the nutrient content in the leaves of plants on prairie dog towns. Thus prairie dogs create strong bottom-up effects on the grassland ecosystem through their grazing and burrowing activities.

This effect does not go unnoticed by larger herbivores such as bison *(Bison bison)* and pronghorn *(Antilocapra americana)* that respond to these increased nutrient levels by preferentially grazing on prairie dog towns (Coppock et al. 1983b; Kreuger 1986). A study of bison and pronghorns grazing at Wind Cave National Park, South Dakota, found that pronghorns spent between 55% and 85% of their time feeding on prairie dog colonies, and between 57% and 97% of that time they were feeding in the forb-dominated centers of the colonies, while bison spent most of their time feeding on the grass-dominated edges of colonies (Kreuger 1986). Another study at Wind Cave found that bison spent 90% of their feeding time on a prairie dog town during the growing season, even though the town represented only 39% of the available habitat in the valley where the study was conducted (Coppock et al. 1983b). Historically, bison grazed in the Great Plains in large numbers, estimated at 30–60 million before 1800 (Fahnestock and Detling 2002), and influenced the development and maintenance of grasslands (Axelrod 1985). Like the prairie dogs, bison graze on grass (Van Vuren and Bray 1983), and grazing affects the cycling of nitrogen in plant communities (Holland and Detling 1990; Green and Detling 2000; Fahnestock and Detling 2002), both through clipping vegetation and through the deposition of feces and urine by the grazing animals (Day and Detling 1990a, b). One study found that although bison urine deposition made up patches that represented only 2% of the available habitat, these patches were responsible for 7% of the total biomass and 14% of the total nitrogen consumed by herbivores in that habitat (Day and Detling 1990a). Because prairie dogs concentrate their grazing activities within the confines of their towns, the colonies represent patches that have higher nutrients than the surrounding grassland (Fahnestock and Detling 2002).

Soil Effects

Prairie dogs can considerably change the soil composition on prairie dog towns. The effect of only 12 prairie dog burrows with a combined volume of 2.7 m³ is the removal of 3.63 metric tons of soil to the ground surface (Koford 1958). The average density of prairie dog burrows varies greatly per town, but if we use a figure of 20 burrows per hectare (2.5 acres) we can calculate that about 600 tons of soils are cycled to the surface for the establishment of a 100-hectare (250-acre) town. Large prairie dog complexes can include as many as 4,000 hectares (10,000 acres) (Van Putten and Miller 1999). One estimate is that prairie dogs can remove about 225 kilograms (500 pounds) of soil per burrow, and densities of 20–40 burrows per hectare contribute to a large amount of soil turnover (Detling and Whicker 1987). This turn-over results in an increase in soil porosity and an increase in organic materials incorporated into the soil (Munn 1993). The increased porosity leads to higher water penetration and groundwater recharge.

Prairie dogs increase the benefit of soils to plants and soil organisms by adding organic matter and nutrient salts to soils, improving soil structure, and increasing water filtration (Koford 1958). One study found that soil moisture was significantly higher on a heavily grazed prairie dog town, even though the soil temperature at a depth of 15 centimeters was 2.7°C greater than in ungrazed areas (Archer and Detling 1986). However, prairie dogs can also compact soil near burrows, increase surface soil calcification by moving calcium-rich ($CaCO_3$) or salt-rich subsoils to the surface and accelerate erosion by increasing the amount of bare ground in an area (Koford 1958; Munn 1993). These latter effects can make the soil less hospitable to plant and invertebrate life. The effect that prairie dogs have on soil structure and nutrient availability is dependent on the unique soil structure of each prairie dog town. In towns with shallow, nutrient poor, or high salinity soils, subsurface soil mixing by prairie dogs tends to increase plant biomass. On towns with good surface soil, but underlying salty or calcareous soil, prairie dog burrowing tends to decrease plant biomass by moving saline soils to the surface (Munn 1993).

Effects of prairie dogs on soils may also differ by species of prairie dog. For example, black-tailed and Mexican prairie dogs have stronger effects on soil composition than the other three species because they tend to excavate larger and more complicated burrow systems. Mexican and black-tailed prairie dogs also mix subsurface soil excavated from their burrows with surface soil to build large, compact mounds. White-tailed, Gunnison's, and Utah species have smaller mounds that consist only of excavated soils and are not purposefully compacted (Smith 1967). Non-compacted mounds support a high diversity of forbs that are attracted to the disturbed soil. However, seeds are not easily established on compacted mounds and they are consequently left bare of vegetation (Cincotta et al.1989).

There is some uncertainty as to whether the effect of prairie dogs on soil is confined to the immediate areas of alteration—burrows and mounds—or if it is more widespread throughout the colony. One study found that black-tailed prairie dog mounds were enriched with phosphorous and had a higher pH but that surrounding soils did not exhibit the same effect (Carlson and White 1987). However, other studies have reported more widespread effects of increased organic matter, nitrogen, and phosphorous throughout prairie dog towns compared to surrounding unoccupied areas (Schloemer 1991; Cincotta et al.1985; Coppock et al. 1983a).

Vegetation Effects

Through grazing and burrowing activities, prairie dogs change the structure of individual plants as well as plant composition on the entire town. The effects vary considerably depending on the time that has elapsed since the site was colonized, on the original colonization conditions, on whether there is substantial large herbivore grazing pressure coupled with prairie dogs, and on what the vegetation in the area surrounding the colony is like. Some differences in plant structure and composition can be detected between different species of prairie dogs as well.

By grazing, prairie dogs can affect the rate of nitrogen uptake in plants, leading to higher levels of nitrogen in grazed plants (Polley and Detling 1989; Holland and Detling 1990; Fahnestock and Detling 2002). One study at Wind Cave National Park, South Dakota, found that mean nitrogen levels in grass leaves declined significantly after prairie dogs were removed from the site (Cid et al. 1991). An experimental study of defoliation of grasses at Wind Cave found that nitrogen concentrations were almost double those of control plots when grasses were clipped weekly, bimonthly, or monthly, simulating grazing by herbivores (Green and Detling 2000). Grazing can also affect the evapotranspiration of grasses, leading to a warmer canopy microclimate with a higher evaporative demand (Day and Detling 1994). The increased nitrogen levels in turn can affect preferential feeding by large ungulates such as bison and pronghorns on plants within prairie dog colonies (Coppock et al. 1983a; Kreuger 1986; Wydeven and Dahlgren 1985). Bison tend to feed on the outskirts of prairie dog towns where the forage is most tender and packed with nutrients. Individual plants that have grazing resistant forms tend to do well on prairie dog towns (Detling and Painter 1983; Painter et al. 1993). Forms of these plants change over time becoming shorter, denser, more compact, and having shoots packed with higher densities of nutrients (Coppock et al. 1983a). Selection pressure from herbivores, such as prairie dogs, may lead to the development of genetically distinct populations of grasses that differ in their morphology and physiology, depending on whether they grow on colonies or in surrounding grassland (Jaramillo and Detling 1988).

Grazing by prairie dogs can also change the species composition and species richness of plants (Hansen and Gold 1977; Bonham and Lewrick 1976; Clark et al. 1982; O'Meilia et al. 1982; Coppock at al. 1983b; Brizuela et al. 1986; Archer et al. 1987; Cincotta et al. 1989; Weltzin et al. 1997 a, b). Most studies on changing plant composition have been conducted on black-tailed prairie dog towns. Generally, taller grasses disappear first as they are most preferred by prairie dogs and least resistant to grazing pressure. Following the disappearance of taller grasses there is a surge in the population of short-grasses and over time these largely give way to annual forbs and dwarf shrubs.

In addition to changes through time, at any given time there are usually distinct zones within a colony that exhibit highly different characteristic plant species. An inner zone, which usually has been colonized the longest, tends to be dominated by forbs, annuals, and shrubs. A transition, or newly colonized zone, surrounds the inner zone and consists of a diversity of perennial grasses, short grasses, and forbs. The boundary of the colony gives way to the plant community that was dominant before colonization (Coppock et al. 1983a; Cincotta et al. 1985). In some instances in the arid Southwest, Gunnison's prairie dog colonies do not seem to exhibit this zoning behavior, perhaps due to the harsh conditions of the landscape or the long, unchanging occupation by prairie dogs (Bangert and Slobodchikoff 2000).

Prairie dogs also influence plant cover through digging for seeds. Such digging can increase the patchiness of grasses on prairie dog colonies (Bangert and Slobodchikoff 2000). Seeds tend to accumulate in depressions around clumps of vegetation (Reichman 1984). Prairie dogs seem to do most of their seed foraging at the base of existing clumps of grass where seeds are predicted to aggregate, creating patches of bare ground.

Besides altering plant composition, prairie dogs act as stabilizers of the grassland ecosystem. Prairie dogs prevent tree and shrub encroachment into grasslands by eating seeds and seedlings of trees—the foraging activities of prairie dogs maintain southwestern grassland ecosystems by preventing the establishment of woody species, such as honey mesquite (Weltzin et al. 1997a). In addition, prairie dog towns may encourage native plant growth. A greater number of native plants grow on Gunnison's prairie dog towns than in areas that are not colonized by prairie dogs, perhaps because native plants are more resistant to grazing pressures from prairie dogs (Slobodchikoff et al. 1988).

Different species of prairie dogs seem to have slightly different effects on vegetation. The ecology of the black-tailed prairie dog is the most studied, and most general trends attributed to prairie dogs come from research on this species. However, black-tailed prairie dogs have quite different behaviors and live in quite different prairie systems than the white-tailed group of the species; for example, they usually live at much higher densities (Clark 1977; Hoogland 1995). Therefore caution should be used when applying literature on black-tailed prairie dogs to the other four species. Some differences between species and the subsequent difference in effects on vegetation follow.

Black-tailed prairie dogs tend to clip grasses to a low height to increase visibility across the prairie dog town (Tileston and Lechleitner 1966; Hoogland 1979; Coppock et al. 1983a). This often results in decreased diversity of vegetation both within and between towns as only certain plants can handle stringent, long-term clipping (Tileston and Lechleitner 1966; Hoogland 1979; Coppock et al. 1983a). In contrast, white-tailed, Gunnison's, and Utah prairie dogs do not actively clip vegetation. This results in higher intra-town diversity in vegetation and blurrier boundaries between towns and surrounding vegetation types (Menkens 1987).

Gunnison's, white-tailed, and Utah prairie dogs also tend to inhabit areas that have steeper slopes and more topographical diversity than black-tailed prairie dog towns (Menkens 1987; Tileston and Lechleitner 1966; Wagner and Drickamer 2004). Topographical diversity on a prairie dog town can also increase diversity of plant species that colonize that town. However, one study reported lower species diversity on Gunnison's prairie dog towns in Arizona in comparison to black-tailed prairie dog towns (Slobodchikoff et al. 1988). Another study of Gunnison's prairie dogs in southern Colorado found no significant differences in biomass and vegetation cover between prairie dog towns and surrounding grassland (Grant-Hoffman and Detling 2006). Densities of white-tailed, Utah, and Gunnison's prairie dog towns are often lower than those of black-tailed prairie dog towns, which could translate into less intensive effects on plant diversity and cover. More studies are needed to determine how variable the effects of different species of prairie dogs are on primary production in grasslands.

Competition between Prairie Dogs and Large Herbivores

Bison, pronghorns, and prairie dogs have coexisted for thousands of years. We can ask the question, to what extent do these herbivores compete with one another? An even more salient question in today's economy is, to what extent do prairie dogs compete with cattle? The question of competition with cattle provides some of the rationale for the historic efforts to eradicate prairie dogs in vast numbers over the last hundred years, contributing to the rapid decline in prairie dog populations. A general perception has been that because prairie dogs and cattle eat grass, they must compete with one another. But finding the truth in this perception is not so easy. Competition is very difficult to document, and if resources are plentiful, there may not be any competition because there are ample resources for everyone. It is only when resources become limited that we begin to see evidence of competition, and even then the evidence is seldom clear-cut.

That seems to be the situation with prairie dogs (Detling 2006). In 1902, C. Hart Merriam, a biologist who was head of the U.S. Biological Survey, suggested that 32 prairie dogs eat as much grass as a sheep, and 256 prairie dogs eat as much grass as a cow (Merriam 1902). Exactly how these numbers were determined is not clear. However, on the basis of these numbers, Merriam calculated that the annual loss of forage from a colony of 400 million prairie dogs in Texas would support 1, 562,000 cattle, and that prairie dogs consume between 50% and 75% of the grass available to cattle, an estimate that served partially as a justification for government efforts to poison prairie dogs. A later estimate by Koford (1958) suggested that 335 prairie dogs would eat as much grass as a single cow, and 70 grams (2.5 ounces) of forage per day would be consumed by a single prairie dog. From this estimate, Detling (2006) calculates that prairie dogs might remove 100% of the average annual net primary production (ANNP, a measure of plant productivity) in least-productive shortgrass steppe colonies and 10% of the ANNP in most-productive mixed-grass colonies, given the assumptions that prairie dog densities are 25 adults and yearlings per hectare (2.5 acres) and the prairie dogs consume about 625 kilograms (1375 pounds) of forage per hectare per year. Detling (2006) cautions, however, that prairie dog densities and the amount of food that they consume varies greatly.

Experimental studies of competition between cattle and prairie dogs are sparse. Hansen and Gold (1977) studied the diets of black-tailed prairie dogs, desert cottontail rabbits *(Sylvilagus audubonii),* and cattle at an experimental range near Nunn, Colorado, in different vegetation types. Prairie dogs were transplanted into two new colonies, one at a site where there was native short-grass vegetation and the other into a revegetating area, with an average of 6–7.3 prairie dogs per hectare (2.4–2.9 prairie dogs per acre). A third site was an established colony of prairie dogs, with an average of 1 prairie dog per hectare (0.4 prairie dogs per acre). Cattle grazed in all three sites. Prairie dog diets were composed primarily of sedges (*Carex* spp.)(36%), blue grama grass *(Bouteloua gracilis)* (20%), sand dropseed *(Sporobolus cryptandrus)* (13%), fringed sagewort *(Artemisia frigida)* (8%), and scarlet globemallow *(Sphaeralcea coccinea)* (7%). Cattle diets were composed primarily of western wheatgrass *(Agropyron smithii)* (26%), sedges (23%), blue grama grass (10%), sand dropseed (10%), needle-and-thread *(Stipa comata)* (6%), scarlet globemallow (6%), and buffalograss *(Buchloe dactyloides)* (2%). The similarity in the diets ranged from a low of 42% in winter to a high of 69% in spring. However, there was no indication about how the cattle might have been affected by the presence of the prairie dogs, in either a positive or negative way. A big problem in trying to ascertain the dietary overlap of prairie dogs and cattle is that prairie dogs are confined to eating the plants that grow in their colonies, while cattle are confined to eating plants that grow in their pastures, so simple measures of dietary overlap might not provide measures of dietary preference.

In another study, O'Meilia et al. (1982) released a total of 311 black-tailed prairie dogs into six of twelve 2.53-hectare-sized pastures over the course of three years, and allowed the prairie dogs a minimum of two years to become established in these pastures. Cattle were allowed to graze in all 12 pastures. After the prairie dogs became established, O'Meilia et al. (1982) sampled over the course of two years the forage that was utilized by the prairie dogs and the cattle, and the weight gains of the cattle that grazed on pastures with prairie dogs and on pastures without prairie dogs. The results of the study showed that the weight gains of the cattle that grazed in pastures without prairie dogs were not significantly different from the weight gains of cattle that grazed in pastures that were colonized by the prairie dogs.

A simulation study attempted to model the relationship between prairie dogs, available forage, and the effects of prairie dog grazing on cattle. Uresk and Paulson (1988) developed a model based on data collected from a 2,100-hectare (5,250-acre) pasture in the Conata Basin near Wall, South Dakota. Included in the model were data from cattle diet composition, black-tailed prairie dog diet composition, forage production, prairie dog forage consumption, and cattle forage consumption, for 44 prairie dogs per hectare. The model varied forage utilization at four levels, 20%, 40%, 60%, and 80%, with the assumptions that both prairie dogs and cattle were grazing on the forage, that adequate amounts of forage were available, and that grazing was proportional to the numbers of herbivores. A further assumption was that prairie dogs occupied only a maximum of 40 hectares (100 acres) of the pasture. The results of the model showed that at 20% of forage utilization, the difference between 0 hectares occupied by prairie dogs and 40 hectares of prairie dogs is 5 cattle (55 cattle at 0 hectares of prairie dogs, 50 cattle at 40 hectares). At 80% forage utilization, the difference between 0 hectares and 40 hectares of prairie dogs is 7 cattle (221 cattle at 0 hectares, 214 cattle at 40 hectares). These are differences on the order of 3.2%–10%, between the number of cattle that the pasture can support with no prairie dogs present and the number of cattle that the pasture can support with 40 hectares of prairie dogs.

So the answer to the question of how much do prairie dogs compete with cattle appears to be, not that much. Under some circumstances, such as abundant resources, there seems to be little, if any, competition. Under other circumstances, such as heavier grazing regimes, there may be some competition, but the competition appears to be relatively slight and monetary yields from cattle appear to be affected relatively little by having the cattle graze in areas where prairie dogs are found. For example, a recent study found that when prairie dogs occupied a pasture, the monetary loss per steer was in the range of 5%–14%, depending on the amount of occupancy of the pasture by prairie dogs, with a low level of occupancy of 20% resulting in a 5% monetary loss per steer, and a high level of occupancy of 60% resulting in a 14% monetary loss per steer (Derner et al. 2006).

Prairie Dogs and the Invertebrate Community

Prairie dogs and prairie dog towns influence invertebrate community structure in three primary ways. First, the burrow environment creates a refuge for many invertebrates seeking to escape from the often harsh, dry conditions in grasslands. Second, by manipulating plant species composition on the colony, prairie dogs also manipulate food sources for invertebrates, causing some populations to increase on prairie dog towns and causing declines in other species. And third, by eating vegetation, prairie dogs can create bare areas that allow greater ease of movement for terrestrial invertebrates.

Temperature fluctuations on North American grasslands are often extreme. They can range from well over 38°C (100°F) on summer days to well below 18°C (0°F) during winter nights. Prairie grasslands are also very arid, with low humidity and steady dry winds. For many invertebrate species these dry conditions and strong fluctuations in temperature create an uncomfortable or inhospitable environment. However, prairie dog burrows are constant environments, with relatively high humidity and steady temperatures (Smith 1982). Burrow systems persist for tens of years to thousands of years (Koford 1958; Carlson and White 1987), and can serve as potential refugia for a variety of arthropod and vertebrate species.

The relationship between prairie dog colonies and invertebrate species remains unclear. The biomass of foliage-dwelling and flying arthropods can decrease on prairie dog towns, possibly due to the higher occurrence of insectivores (O'Meilia et. al. 1982). However, ground-dwelling arthropods (specifically darkling beetles in the family Tenebrionidae) seem to benefit from the equable environments in prairie dog burrows and thus can increase in abundance. In a comparative study of the arthropods found on active Gunnison's prairie dogs towns, abandoned towns, and adjacent grassland at Petrified Forest National Park, Arizona, Bangert and Slobodchikoff (2006) found that even though species richness was almost the same in each habitat, the groups that were present and their abundances differed. In grassland sites, the herbivorous leafhoppers in the family Cicadellidae were 10 times more abundant than on prairie dog colonies, while the dung-feeding Scarabeidae were 13 times more abundant on active prairie dog towns. Arthopods in the groups Pholcidae (pholcid spiders), Mutillidae (velvet ants), Curculionidae (weevils), Cerambycidae (long-horned beetles), and Isopoda (isopods) were primarily found on inactive towns, while the groups Salticidae (jumping spiders), Gelichiidae (moths), and Gryllidae (crickets) were primarily found on active towns. Among the tenebrionid beetles, *Eleodes extricata* was found in larger numbers on inactive towns, usually in close proximity to abandoned prairie dog burrows, while *Eleodes hispilabris* was found in equal numbers on both active and inactive towns, and seemed to prefer any type of burrow.

Prairie dogs may affect some arthropods in species-specific ways. For example, while the harvester ant species, *Pogonomyrmex rugosus*, exists in higher abundance on prairie dog towns than on uncolonized sites, two other species of *Pogonomyrmex* (*P. barbatus, P. occidentalis*) are more abundant on non-colonized sites than on prairie dog towns (Kretzer and Cully 2001). Pollinating species of invertebrates, especially bees and butterflies, may increase on prairie dog towns, possibly due to the greater amount of flowering forbs found on prairie dog towns compared to adjacent uncolonized lands. Heavy grazing of foliage by herbivores in prairie dog towns can reduce root growth below ground and lead to higher population levels of root nematodes (Ingham and Detling 1984).

By creating areas of bare ground around the burrow entrances and within the centers of towns, prairie dogs can facilitate the movement of invertebrates (Bangert and Slobodchikoff 2000, 2004). For example, *Eleodes hispilabris* tenebrionid beetles moved 44% faster, in straighter paths, and 63% farther in Gunnison's prairie dog towns at Petrified Forest National Park, Arizona, than they did in surrounding grasslands, probably because the grasslands provided an impediment to rapid movement (Bangert and Slobodchikoff 2004). The tenebrionid beetles may also derive another benefit from the bare ground areas. Many of these beetles have defensive secretions that are exuded

or squirted out of glands near their anus, and when they are attacked by a
vertebrate predator, they have a headstand behavior that elevates their poste-
rior and maximizes the chances that the predator will be sprayed or affected
by the defensive secretions (Slobodchikoff 1987). The bare ground allows the
beetles to perform this antipredator headstand display, which they might not
be able to do amid thick clumps of grass.

The interactions of prairie dogs and invertebrates are fascinating yet still
not well understood. More research on possible competitive interactions
between grasshoppers, ants, and prairie dogs would be interesting. Burrows
as refuges for invertebrates needs to be better quantified as well as the role of
prairie dog colonies as rich sites for pollinating insects.

Prairie Dog Towns, Reptiles, and Amphibians

Prairie dogs can alter habitats in several ways that could potentially affect
the presence or abundance of small vertebrates. Structural changes such as
burrows provide refugia, hibernacula, nesting, or resting sites for birds, like
burrowing owls (McCracken et al. 1985; O'Meilia et al. 1982), small rodents,
and a variety of amphibians, lizards, and snakes. In addition, vegetation shifts
induced by prairie dogs may alter arthropod abundance and diversity as
discussed above. This may then alter the availability of arthropod prey for
insectivorous rodents, birds, and reptiles.

Burrows act as refugia for a wide variety of snakes and lizards, and rattlesnakes are known to be predators of prairie dogs. Equable burrow environments coupled with availability of both insects and larger rodent prey populations makes prairie dog towns enticing environments for reptiles and some
amphibians, like tiger salamanders and some toads. For example, the lesser
earless lizard, *Holbrookia maculata,* is more abundant on Gunnison's prairie
dog colonies than in the surrounding grassland, and when artificial burrows
were created away from prairie dog colonies, the abundance of the lizards
increased in the vicinity of the artificial burrows (Davis and Theimer 2003).

Birds and Prairie Dog Towns

Prairie dog towns provide habitat, food, nesting, and resting sites for many
different bird species. Insectivorous birds, raptors, and birds that require

burrows or open habitats for nesting seem to benefit the most from prairie dog towns. Burrowing owls, horned larks, western meadowlarks, cliff swallows, mourning doves, mountain plovers, and a wide variety of raptors are seen more on prairie dog towns than on adjacent uncolonized lands (Barko et al. 1999; Desmond and Savidge 1996; Gietzen et al. 1997; Lomolino and Smith 2004).

Some studies suggest that the overall bird species diversity is higher on prairie dog towns than in surrounding grasslands, although some of the species that have been found on the prairie dog towns represent introduced exotics rather than native birds (Agnew et al. 1986; Kotliar et al. 1999). In South Dakota, 29 bird species have been found on black-tailed prairie dog colonies, and in Oklahoma, 83 bird species have been found on 250 black-tailed colonies (Campbell and Clark 1981; Tyler 1968; Barko et al. 1999). In another survey of 36 sites with black-tailed prairie dogs and 36 sites without prairie dogs in the panhandle of Oklahoma, the average species richness of birds (i.e., the number of species) was significantly higher on colonies in the summer, but not in the fall (Smith and Lomolino 2004). Over the course of three years, mean species richness of birds on prairie dog sites was 9.5, and the mean species richness of birds on non-prairie dog sites was 8.2, a significant difference. In the fall, the mean species richnesses were 5.7 for the prairie dog sites and 5.6 for the non-prairie dog sites. In summer, burrowing owls *(Athene cunicularia)*, killdeer *(Charadrius vociferous)*, horned larks *(Eremophilia alpestris)*, and meadowlarks *(Sturnella* spp) were most often found on prairie dog towns, while in fall, horned larks and ferruginous hawks *(Buteo regalis)* were species that tended to occur more often in association with prairie dogs (Smith and Lomolino 2004).

Of all the bird species, burrowing owls and mountain plovers *(Charadrius montanus)* may be most dependent on prairie dog towns, although more research is needed to establish the exact degree of dependence. Mountain plovers selectively use black-tailed prairie dog towns to nest, breed, and feed, and are rarely found outside of prairie dog towns. Large open areas without vegetation that occur more often on prairie dog towns may provide good nesting sites for the mountain plover. An increase in invertebrate prey, especially beetles, on prairie dog towns may influence mountain plover preference for these sites (Knowles et al. 1982; Olson 1984; Olson and Edge 1985).

Burrowing owl densities are greater in prairie dog towns than in areas where other burrows are available. As the size of prairie dog towns increases, more burrowing owls occupy the towns but with less density (Desmond and Savidge 1996). Decline of burrowing owls has been linked to the decline of black-tailed prairie dogs (Desmond et al. 2000). However, there is a significant time lag between the demise of prairie dogs in a town and a decline of burrowing owls, because it takes time for the burrow systems to collapse and fill in after prairie dogs are gone.

Raptor populations are greatly influenced by prairie dog populations. Raptors feed directly on prairie dogs and also on other associated rodents that live on prairie dog towns. Ferruginous hawks, rough-legged hawks, red-tailed hawks, prairie falcons, bald eagles, and golden eagles prey directly on prairie dogs, and their populations decrease when prairie dogs are eliminated and increase when prairie dogs rebound (Gietzen et al. 1997). Declining prairie dog towns particularly affect ferruginous hawks that depend on prairie dogs for a large portion of the winter diet (Jones 1989; Allison et al. 1995).

Rodents and Rabbits

Prairie dog towns affect rodent and rabbit distributions. The general pattern seems to be that species richness of small mammals declines on prairie dog towns, while abundance of small mammals increases (Agnew et al. 1982). In other words, a few species benefit greatly from the existence of prairie dog towns while others are outcompeted by prairie dogs. Insectivorous rodents in particular, such as grasshopper mice, benefit from prairie dog towns. Deer mice *(Peromyscus* spp.*)*, northern grasshopper mice *(Onchymys leucogaster)*, prairie voles *(Microtus ochrogaster)*, 13 lined ground squirrels *(Spermophilus tridecemlineatus)*, western harvest mice *(Perognathus* spp.*)*, and non-native house mice *(Mus musculus)* tend to increase in abundance. One study with Gunnison's prairie dogs found that transplanting the prairie dogs to a new colony did not affect the other rodents at the site over a one-year period (Davidson et al. 1999). However, it is possible that there may be a considerable time lag between colonization of prairie dogs and effects on other rodents.

Prairie Dog Predators

Prairie dogs are common prey for many grassland predators as well as occasional grassland visitors. The long history of prairie dogs as prey animals is

Figure 6.1. Black-footed ferret, an endangered species. The principal food of the black-footed ferret is prairie dogs. (Photo by C. N. Slobodchikoff.)

seen most clearly in the evolution of their sophisticated alarm call system. Prairie dogs have such a dominant role as prey species that they have adapted sophisticated alarm call systems to warn of approaching intruders. These alarm call systems carry information about different predator types and may cause different defense or flee responses. The calls may also be adapted to the vegetation structure on the colony allowing the vocalizations to carry optimally through the environment so that prairie dogs are adequately warned (Perla and Slobodchikoff 2002).

The most common predators of prairie dogs are red-tailed and ferruginous hawks, golden eagles, coyotes, badgers, and black-footed ferrets (in historical population levels). Predators range in size from bears and cougars to black-footed ferrets that are about the same size as a full-grown prairie dog. Because of their stationary nature, prairie dog towns provide reliable long-term sources of food for predators. This is a fairly unique opportunity for predators to have such a concentrated and stable food source.

The predator most dependent on prairie dog complexes for continued survival is the black-footed ferret *(Mustela nigripes)* (Figure 6.1). As much as 95% of the black-footed ferret diet comes from prairie dogs. In addition, the black-footed ferret requires abandoned prairie dog burrows for shelter and breeding. The historic range of the black-footed ferret corresponds directly to prairie dogs and the decline of the black-footed ferret has been directly linked to the decline of prairie dogs (Hall 1951; Linder 1973; Clark 1989).

Ferrets move an average of 125 meters (375 feet) per night in search of prey (Polderboer et al. 1941). They have large home ranges and require a substantial number of prairie dogs for sustenance. They tend to not occur on any prairie dog towns less than 6 hectares (15 acres) in size and are far more numerous in prairie dog towns that have multiple burrow entrances, a sign that the prairie dog towns are large and well-established (Hillman 1968; Sheets et al. 1971). It is estimated that from 412 to 1,412 prairie dogs are needed to sustain a female ferret and her young for one year (Stromberg et al. 1983). For these reasons, black-footed ferrets require large complexes of prairie dog towns to survive, a rare resource in today's grasslands.

Species That Depend on Prairie Dogs

Prairie dogs and their activities influence many grassland animals. However, at least five species are largely dependent on prairie dogs and prairie dog towns for their continued survival (Knowles and Knowles 1994; Miller et al. 1994). As mentioned above, the endangered black-footed ferret depends on prairie dogs for food and prairie dog burrows for a home. This species' decline is directly linked to the disappearance of large prairie dog complexes. Although federal and state agencies have attempted to reintroduce the black-footed ferret into the wild, these reintroduction efforts have largely failed, with introduced populations declining or disappearing. The only location where black-footed ferrets have been established and are increasing is in the Conata Basin at Buffalo Gap National Grassland, South Dakota, where the ferrets were introduced in 1996 (Kotliar et al. 2006). However, there has been large-scale poisoning of prairie dogs in the Conata Basin since the black-tailed prairie dog was removed from the candidate list of threatened and endangered species, so the future of the black-footed ferret population there is subject to some question. The rare and threatened swift fox *(Vulpes velox)* also depends on the prairie dog for habitat and food; one study of the swift fox showed that prairie dogs made up 49% of its diet (Uresk and Sharps 1986). Burrowing owls and mountain plovers both nest on prairie dog towns. The mountain plover is an insectivore that depends on the great abundance of tenebrionid beetles and other arthropods that are more abundant on prairie dog towns than lands unoccupied by prairie dogs. The burrowing owl likely gets sustenance from the high density of small mice, voles, and rats that live on prairie dog towns. Prairie dogs make up a large part of the winter diet of ferruginous hawks, whose numbers are declining rapidly. Without attention to waning prairie dog populations, these species will likely decline and possibly go extinct.

Prairie Dog Colonies as Islands

Although in the past the colonies of prairie dogs were large and covered thousands of hectares, the colonies of today are relatively small and in many cases represent tiny patches across the landscape of grassland, agriculture, and urban development. These patches can be considered as islands in a sea of grasslands and other habitats. In their seminal book on island biogeography, MacArthur and Wilson (1967) outlined some predictions for relationships between species numbers (species richness) and island area. A principal prediction of the theory of island biogeography is that species richness should increase with the size of the island. For example, Brown (1978) found that mountain ranges in Nevada, existing as isolated islands in a sea of desert, support the species-area prediction of island biogeography, in that the ranges with larger amounts of habitat had greater numbers of mammalian species.

Evidence for prairie dog towns as islands is not abundant. In one study, Lomolino and Smith (2003) sampled 36 black-tailed prairie dog towns and 36 nearby grassland, cropland, and pasture sites over the course of three years in Oklahoma, surveying the towns for the presence of non-flying vertebrates. The prairie dog towns ranged in size from 9 to 211 hectares (22 to 527 acres). The surveys detected 29 species of mammals, 20 species of reptiles, and 5 species of amphibians, with species richness on any one town ranging from 3 to 13 species of mammals, 0 to 5 species of reptiles, and 0 to 3 species of amphibians. In their study, Lomolino and Smith (2003) did not find any significant correlation between the number of species present and the area of the towns, suggesting that the towns did not fit the predictions of island biogeography theory. Instead, species richness was significantly correlated with the larger surrounding landscape habitat, such as grassland, cropland, or forest. On the other hand, the study by Bangert and Slobodchikoff (2006) with arthropods on prairie dog towns in Petrified Forest National Park, Arizona, (see pages 128–130), suggests that species richness is not the most appropriate measure to use in studying whether prairie dog towns can be considered islands. In Bangert and Slobodchikoff's (2006) study, species richness was not significantly different between active prairie dog towns, abandoned towns, and surrounding grassland, but the composition of species and the ecological role that they played differed between these habitats.

Keystone Species of the Great Plains Grassland

The effects of prairie dog modification of grasslands are far-reaching across trophic levels. But how integral is the prairie dog to this system? Is it an equal participant or do the effects of prairie dog activities far outweigh what you would expect from a single species? Most authors believe that prairie dogs have a disproportionately large effect on the prairie ecosystem and they are willing to call the prairie dog a keystone species (Knowles and Knowles 1994; Miller et al. 1994). However, a few are cautious with the use of this term for prairie dogs and call for more careful research into this topic (Stapp 1999).

For many years, ecologists' prevailing view of ecosystem structure was one of top-down forces—for example, predators controlling populations of animals lower on the food chain, such as herbivores (Hairston et al. 1960). This view was evident in the formulation of the keystone species concept, where the keystone species was a predator having a large effect on community structure (Paine 1969). More recently, a keystone species has been defined as one whose overall impact on a community is disproportionately large with respect to its abundance (Power et al. 1996). A somewhat more specific definition of keystone species is (Miller et al. 2000): "Keystone species influence ecosystem structure, composition, and function in a unique and significant manner through their activities, and the effect is disproportionate to their numerical abundance." Kotliar (2000) added the stipulation that the species must perform a function that is largely unduplicated by other species.

In addition to the keystone species concept, there is also the concept of ecosystem engineers, organisms that modify their habitat in disproportionately large ways. An ecosystem engineer can be a keystone species as well. A familiar example is the beaver who can make large-scale modifications to rivers and streams. Prairie dogs, with their burrow systems, are another example

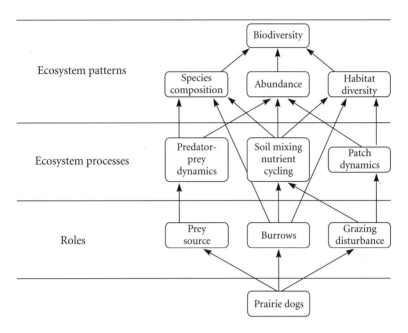

Figure 6.2. Different pathways by which prairie dogs affect ecosystem processes. (From
  Kotliar et al. 1999, with kind permission of Springer Science and Business
  Media.)

of an ecosystem engineer (Bangert and Slobodchikoff 2000). In contrast to
the original keystone species definition, ecosystem engineers are bottom-up
forces (such as herbivores, that are lower on the food chain than predators)
that modify landscapes and potentially modify communities (Jones et al. 1994;
Lawton 1994). In one marine example of the intersection between the keystone
species concept and the ecosystem engineer concept, a keystone species (the
sea otter) influences an ecosystem engineer (the sea urchin) that acts upon
another ecosystem engineer (kelp), resulting in community collapse through
predominately bottom-up forces (Jones et al. 1994). However, in this case, the
ecosystem engineers (urchins and kelp) are not considered keystone species
because they modify their environment in proportion to their abundance.
Recently, other authors have advocated the view that top-down and bottom-
up forces act on communities simultaneously, but that bottom-up forces are
primary (Hunter and Price 1992). Other authors argue that keystone species
can be defined by many different criteria including "keystone modifiers" where

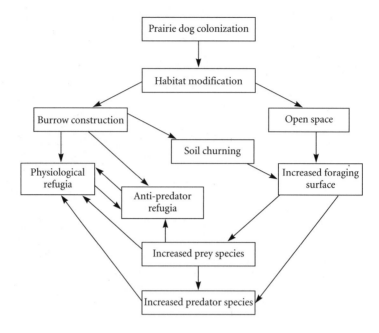

Figure 6.3. Ecological interactions that result from prairie dog colonization.

keystone species can have an inordinate effect on community structure through ecosystem engineering (Mills et al. 1993).

Prairie Dogs and the Grassland Ecosystem

There is little doubt that prairie dogs have a strong effect on their grassland ecosystem (Figure 6.2). Initially, research on species associations with prairie dog towns was anecdotal and observational. However, recent experimental studies are more profoundly uncovering the role of prairie dogs in the grassland system. While some scientists still remain cautious of the keystone term, it is clear that prairie dogs play a strong, influential role in the functioning of North American grasslands and their influence is mainly through bottom-up forces. Prairie dogs affect the species diversity and species composition of plants on colonies, influence nutrient cycling, improve water percolation and soil mixing, and provide a habitat for numerous other species of animals, both vertebrates and invertebrates (Figure 6.3). Their existence contributes to a number of processes that keep the grassland ecosystem both healthy and diverse.

Interlude:
Vertebrate Species Associated with Prairie Dogs

By Richard Reading

Updated from Kotliar et al. (1999).

Species	Benefit[a]	Citation[b]
Obligate associate (N = 1)		
Black-footed ferret, *Mustela nigripes*	PD, BU	1, 8, 38
Strong, well documented benefits of association (N = 16)		
Tiger Salamander, *Ambstoma tigrinum*	BU, TP	2, 3, 13, 14, 25, 27, 28
Lesser earless lizard, *Holbrookia maculata*	BU, PV	2, 4, 13, 30
Prairie rattlesnake, *Crotalus viridus*	BU, PD, OP	13, 14, 27, 28, 32, 33
Mountain plover, *Charadrius montanus*	OV	2, 3, 7, 11, 26, 34
Killdeer, *Charadrius vociferous*	OV	2, 3, 6, 7, 8, 10, 28, 33, 34
Golden eagle, *Aquila chrysaetos*	PD, OP	1, 2, 3, 7, 8, 33
Ferruginus hawk, *Buteo regalis*	PD, OP	2, 3, 4, 6, 7, 8, 11, 21, 28, 29, 31, 33, 36
Burrowing owl, *Athene cunicularia*	BU, OV	2, 3, 4, 6, 7, 8, 11, 17, 18, 19, 20, 22, 28, 33, 34, 35, 36
Horned lark, *Eremophila alpestris*	OV	2, 3, 4, 6, 7, 8, 10, 11, 28, 33, 34, 36
Eastern cottontail, *Sylvilagus floridanus*	BU, PV	8, 27, 28, 33
Desert cottontail, *Sylvilagus audubonii*	BU, PV	1, 2, 3, 4, 8, 9, 12, 14, 33, 37
N. grasshopper mouse, *Onychomys leucogaster*	PV	1, 3, 6, 8, 27, 28, 33
Coyote, *Canis latrans*	BU, PD, OP	1, 2, 3, 4, 7, 8, 12, 27, 28, 33
Badger, *Taxidea taxus*	BU, PD, OP	1, 2, 3, 4, 7, 8, 12, 27, 28, 33
Swift fox, *Vulpes velox*	PD, OP	2, 8, 27, 28, 33
Pronghorn, *Antilocapra americana*	PV	1, 3, 4, 7, 8, 27, 28, 33
Some documented benefits of association (N = 38)		
Plains spadefoot toad, *Spea bombifrons*	BU, TP	2, 13, 14, 27, 28
Woodhouse's toad, *Bufo woodhousii*	BU, TP	2, 13, 27, 28

Species	Benefit[a]	Citation[b]
Great plains toad, *Bufo cognatus*	BU, TP	8, 27, 28
Ornate box turtle, *Terrapene ornate*	BU, PV	2, 4, 13, 27, 28
Texas horned lizard, *Phrynosoma cornutum*	BU, PV	2, 4, 13, 27, 28
Little striped whiptail, *Cnemidophorus inornatus*	BU, PV	4, 30
*Plateau striped whiptail, *Cnemidophorus velox*	BU, PV	30
Plains garter snake, *Thamnophis radix*	BU, OP	8, 14, 27, 28
Long-billed curlew, *Numenius americanus*	PV, TP, OV	2, 6, 7, 8, 10, 11, 33
*Aplomado falcon, *Falco femoralis*	PD, OP, OV	16
American kestrel, *Falco sparverius*	OP	2, 3, 4, 6, 7, 8, 9, 10, 33
Swainson's hawk, *Buteo swainsoni*	PD, OP	2, 3, 4, 6, 7, 8, 9, 10, 11
Red-tailed hawk, *Buteo jamaicensis*	PD, OP	1, 2, 3, 6, 8, 9, 33
Scaled quail, *Callipepla squamata*	PV	2, 10, 33
Scissor-tailed flycatcher, *Tyrannus forficatus*	PV	2, 28, 33
Loggerhead shrike, *Lanius ludovicianus*	PV, OP	2, 3, 6, 7, 8, 10, 28, 33
Barn swallow, *Hirundo rustica*	PV	2, 3, 6, 7, 8, 10
Lark sparrow, *Chondestes grammacus*	PV	2, 4, 7, 8, 10, 11, 28, 33
Eastern meadowlark, *Sturnella magna*	PV	2, 28, 33
Brown-headed cowbird, *Molothrus ater*	PV	2, 3, 7, 8, 28, 33
Chestnut-collared longspur, *Calcarius ornatus*	OV	2, 3, 6, 7, 8, 28, 33, 36
*Banner-tailed kangaroo rat, *Dipodomys* spectabilis*	BU, OV	12, 23
*Merriam's kangaroo rat, *Dipodomys merriami*	BU, OV	12
Deer mouse, *Peromyscus maniculatus*	PV	1, 2, 3, 6, 8
W. harvest mouse, *Reithrodontomys megalotis*	BU, PV	6, 8, 12
Silky pocket mouse, *Perognathus flavus*	BU, PV	3, 8, 9, 12
Hispid pocket mouse, *Chaetodipous hispidus*	PV	6, 8, 12
*Tawny-bellied cotton rat, *Sigmodon fulviventer*	PV	12
*White-throated woodrat, *Neotoma albigula*	BU, OV	12
*S. grasshopper mouse, *Onychomys torridus*	BU, OV	12
Mearn's grasshop. mouse, *Onychomys arenicola*	PV	9
Spotted ground squirrel, *Spermophilus spilosoma*	BU, PV	2, 12
Thirteen-lined ground squirrel, *Spermophilus tridecemlineatus*	BU, PV	1, 2, 3, 4, 6, 8, 27, 28, 33
Long-tailed weasel, *Mustela frenata*	BU, PD, OP	3, 4, 8, 12
*Kit fox, *Vulpes macrotis*	BU, PD, OP	12, 15
Spotted skunk, *Spilogale putorius*	OP	8, 12
*Hog-nosed skunk, *Conepatus mesoleucus*	BU, OP	12
White-tailed deer, *Odocoileus virginianus*	PV	8, 27, 28

Mixed or inconclusive evidence of a beneficial association (N = 13)

| Chorus frog, *Pseudacris triseriata* | BU, TP | 3, 8, 14 |

Species	Benefit[a]	Citation[b]
Gopher snake, *Pituophis melanoleucus catenifer*	BU, PV	2, 3, 13
*Great Plains skink, *Eumeces obsoletus*	BU, PV	13
*Wandering garter snake, *Thamnophis elegans*	BU, OP	14
Northern harrier, *Circus cyaneus*	OP	1, 2, 3, 6, 7, 8, 11, 33
Mourning dove, *Zenaida macroura*	OV	2, 3, 4, 6, 7, 8, 33, 34
Lark bunting, *Calamospiza melanocorys*	PV	2, 3, 6, 7, 10, 11, 34, 36
McCown's longspur, *Calcarius lapponicus*	OV	2, 3, 7, 8, 28, 33
Western meadowlark, *Sturnella neglecta*	PV	2, 3, 4, 6, 7, 8, 10, 28, 33, 34
Common grackle, *Quiscalus quiscula*	PV	6, 10
White-footed mouse, *Peromyscus leucopus*	PV	5, 23
Black-tailed jackrabbit, *Lepus californicus*	PV	1, 2, 4, 8, 9, 27, 28
Striped skunk, *Mephitis mephitis*	BU, OP	2, 3, 8, 12, 27, 28

No, or insufficient, evidence of a beneficial association (N = 105)

Couch's spadefoot toad, *Scaphiopus couchi*	TP	2
Green toad, *Bufo debilis*	TP	2
Texas toad, *Bufo speciosus*	TP	2
Gr. plains narrow-mouthed frog, *Gastrophryne olivacia*	BU, TP	2
Yellow mud turtle, *Kinosternon flavescens*	BU, TP	2, 12
Bullsnake, *Pituophis melanoleucus sayi*	BU, OP	14
Western diamondback rattler, *Crotalus atrox*	PD, OP, BU	1, 2
Mohave rattlesnake, *Crotalus scutulatus*	PD, OP, BU	9
*Black-tailed rattlesnake, *Crotalus molossus*	BU, PD, OP	12
*Lined snake, *Tropidoclonion lineatum*	BU, OP	14
*Mexican garter snake, *Thamnophis eques*	BU	12
*Milk snake, *Lampropeltis triangulum*	BU, OP	13
Eastern fence lizard, *Sceloporus undulates*	BU	2
Short horned lizard, *Phrynosoma douglassi*	BU, PV	3, 14
6-lined racerunner, *Cnemidophorus sexlineatus*	BU, OP	2, 13
Texas spotted whiptail, *Cnemidophorus qularis*	BU, PV	2
*Desert grassland whiptail, *Cnemidophorus uniparens*	BU	12
Chihuahua spotted whiptail, *Cnemidophorus exsanguis*	BU	12
*Desert grassland whiptail, *Cnemidophorus uniparens*	BU	12
*Many-lined skink, *Eumeces multivirgatus*	BU, PV	14
American avocet, *Recurvirostra americana*	TP	7

Species	Benefit[a]	Citation[b]
Lesser golden plover, *Pluvialis dominica*	OV, TP	2
Marbled godwit, *Limosa fedoa*	TP	7
Willet, *Catoptrophorus semipalmatus*	TP	7
Greater yellowlegs, *Tringa melanoleuca*	TP	7
Lesser yellowlegs, *Tringa flavipes*	TP	7
Wilson's phalarope, *Phalaropus tricolor*	TP	7, 8
Long-billed dowitcher, *Limnodromus scolopaceus*	TP	8
Bairds's sandpiper, *Calidris bairdii*	OV, TP	2
Buff-breasted sandpiper, *Tryngites subruficollis*	OV, TP	2
*Whimbrel, *Numenius phaeopus*	OV, PV	11
*Greater roadrunner, *Geococcyx californianus*	OP	11
Turkey vulture, *Cathartes aura*	PD, OP	2, 6, 8, 10
Prairie falcon, *Falco mexicanus*	OP	1, 2, 3, 6, 7, 8, 11
Bald eagle, *Haliaeetus leucocephalus*	PD, OP	2, 8
Mississippi kite, *Ictinia mississippiensis*	OP	2, 10
Sharp-shinned hawk, *Accipiter striatus*	OP	7
*Zone-tailed hawk, *Buteo albonotatus*	OP	11
Rough-legged hawk, *Buteo lagopus*	PD, OP	1, 2, 8, 33
Crested caracara, *Polyborus plancus*	PD, OP	9
Merlin, *Falco columbarius*	OP	7, 8, 10
Lesser prairie chicken, *Tympanuchus pallidicinctus*	OV	2
Sharp-tailed grouse, *Tympanuchus phasianellus*	OV	1, 6, 7, 8
Sage grouse, *Centrocerus urophasianus*	OV	3, 7
Great horned owl, *Bubo virginianus*	OP	3, 6
Snowy owl, *Nyctea scandiaca*	PD, OP	8
*Short-eared owl, *Asio flammeus*	PV	11
Northern flicker, *Colaptes auratus*	PV	2, 7, 8
Eastern kingbird, *Tyrannus tyrannus*	PV	2, 6, 7, 8, 10
Western kingbird, *Tyrannus verticalis*	PV	2, 3, 6, 7, 8, 10
Cassin's kingbird, *Tyrannus vociferans*	PV	10
*Ash-throated flycatcher, *Myiarchus cinerascens*	PV	11
Say's phoebe, *Sayornis saya*	PV	6, 8
Violet-green swallow, *Tachycineta thalassina*	PV	8
N. rough-winged swallow, *Stelgidopteryx serripennis*	PV	6, 8
Cliff swallow, *Hirundo pyrrhonota*	PV	2, 3, 7, 8, 10
*Tree swallow, *Tachycineta bicolor*	PV	11
*Bank swallow, *Riparia riparia*	PV	11
Black-billed magpie, *Pica pica*	PV	3, 7, 8
Chihuahuan raven, *Corvus cryptoleucus*	PV	2, 10
American crow, *Corvus brachyrhynchos*	OP	2, 6, 8, 10
Common raven, *Corvus corax*	OP	8
American robin, *Turdus migratorius*	PV	7, 8
Northern shrike, *Lanius excubitor*	OP	8

Species	Benefit[a]	Citation[b]
Sage thrasher, *Oreoscoptes montanus*	PV	7
Curved-billed thrasher, *Toxostoma curvirostre*	PV	4, 10
Water pipit, *Anthus spinoletta*	PV	8
Sprague's pipit, *Anthus spragueii*	PV	2
Rufous-sided towhee, *Pipilo erythrophthalmus*	PV	8
Vesper sparrow, *Pooecetes gramineus*	PV	2, 3, 7, 8, 10, 11
Savannah sparrow, *Passerculus sandwichensis*	PV	2, 3, 7
Chipping sparrow, *Spizella passerine*	PV	8
*Song sparrow, *Melospiza melodia*	PV	11
*Black-throated sparrow, *Amphispiza bilineata*	PV	11
Slate-colored junco, *Junco hyemalis*	PV	8
White-crowned sparrow, *Zonotrichia leucophrys*	PV	8
Lapland longspur, *Calcarius lapponicus*	PV	2
Snow buntings, *Plectrophenax nimalis*	PV	3
Dickcissel *Spiza americana*	PV	8
Bobolink, *dolichonyx oryzivorus*	PV	8
*Cactus wren, *Campylorhynchus brunneicapillus*	PV	11
*Rock wren, *Salpinctes obsoletus*	BU	24
*Lucy's warbler, *Vermivora luciae*	PV	11
*Yellow-rumped warbler, *Dendroica coronata*	PV	11
Brewer's blackbird, *Euphagus cyanocephalus*	PV	2, 7, 8
Yellow-headed blackbird, *Xanthocephalus xanthocephalus*	PV	6, 7, 8, 11
Boat-tailed grackle, *Quiscalus major*	PV	2
Pine siskin, *Carduelis pinus*	PV	8
American goldfinch, *Carduelis tristis*	PV	8
*House finch, *Carpodacus mexicanus*	PV	11
*Blue grosbeak, *Guiraca caerulea*	PV	11
Eastern mole, *Scalopus aquaticus*	PV	2
White-tailed jackrabbit, *Lepus townsendii*	PV	1, 3, 7, 8
Least chipmunk, *Eutamias minimus*	PV	3
Northern pocket gopher, *Thomomys talpoides*	PV	8
Plains pocket gopher, *Geomys bursarius*	PV	2, 4, 8
Southern plains woodrat, *Neotoma micropus*	PV	2, 4
Richardson's ground squirrel, *Spermophilus richardsonii*	BU, PV	7
Elk, *Cervus elaphus*	PV	7
Mule deer, *Odocoileus hemionus*	PV	3, 7, 8
Bison, *Bison bison*	PV	1, 2, 8
Red fox, *Vulpes fulva*	OP	3, 7, 8
Bobcat, *Lynx rufus*	PD, OP	1, 2, 8
*Ringtail, *Bassariscus astutus*	OP	12

Species	Benefit[a]	Citation[b]
At least some documented evidence of a negative association (N = 16)		
*Yellow-bellied racer, *Coluber constrictor*	BU, OP	13
Upland sandpiper, *Bartramia longicauda*	PV	2, 6, 7, 8, 33
Northern bobwhite, *Colinus virginianus*	PV	2, 33
Common nighthawk, *Chordeiles minor*	PV	2, 3, 7, 10, 33
Grasshopper sparrow, *Ammodramus savannarum*	PV	6, 7, 8, 11, 33, 34, 36
Cassin's sparrow, *Aimophila cassinii*	PV	10, 11, 33, 36
Brewer's sparrow, *Spizella breweri*	PV	7, 11, 36
Baird's sparrow, *Ammodramus bairdii*	PV	7, 36
Northern mockingbird, *Mimus polyglottos*	PV	2, 8, 10, 33
Bullock's oriole, *Icterus bullockii*	—	2, 33
Red-winged blackbird, *Agelaius phoeniceus*	PV	2, 6, 7, 8, 10, 33
Ord's kangaroo rat, *Dipodomys ordi*	PV	1, 2, 4, 8, 9
Plains harvest mouse, *Reithrodontomys montanus*	PV	8
Prairie vole, *Microtus ochrogaster*	PV	6, 8
Raccoon, *Procyon lotor*	—	2, 3, 8, 12
*Hooded skunk, *Mephitis macroura*	BU, OP	12
No data on association, but species life history suggests an accidental occurrence (N = 48)		
Northern leopard frog, *Rana pipiens*		3
Western toad, *Bufo boreas*		3
Bullfrog, *Rana catesbeiana*		8
Sagebrush lizard, *Sceloporus graciosus*		3, 4
Chihuahua spotted whiptail, *Cnemidophorus exsanguis*		4, 12
Smooth green snake, *Opheodrys vernalis*		8
Common garter snake, *Thamnophis sirtalis*		3
Eared grebe, *Podiceps nigricollis*		7
Pied-billed grebe, *Podilymbus podiceps*		7
White pelican, *Pelecanus erythrorhynchos*		7
Double-crested cormorant, *Phalacrocorax auritus*		7
Black-crowned night heron, *Nycticorax nycticorax*		7
Great blue heron, *Ardea herodias*		7, 8
*Osprey, *Pandion haliaetus*		11
Trumpeter swan, *Cyngus buccinator*		8
Snow goose, *Chen caerulescens*		2
Canada goose, *Branta canadensis*		7, 8
Mallard, *Anas platyrhynchos*		6, 7, 8
Gadwall, *Anas strepera*		7, 8
Ruddy duck, *Oxyura jamaicensis*		7
Green-winged teal, *Anas crecca*		3, 7

Species	Benefit[a]	Citation[b]
American wigeon, *Anas americana*		7
Northern pintail, *Anas acuta*		6, 7, 8
Northern shoveler, *Anas clypeata*		8
Blue-winged teal, *Anas discors*		6, 7, 8
Canvasback, *Aythya valisineria*		8
Redhead, *Aythya americana*		7
Lesser scaup, *Aythya affinis*		7
Sora, *Porzuna Carolina*		6, 8
American coot, *Fulicaa americana*		7
Ring-billed gull, *Larus delawarensis*		7, 8
Herring gull, *Larus argentatus*		7
California gull, *Larus californicus*		7
Belted kingfisher, *Ceryle alcyon*		8
Red-headed woodpecker, *Melanerpes erythrocephalus*		8
Downy woodpecker, *Picoides pubescens*		8
Ladder-backed woodpecker, *Picoides scalaris*		2
Blue jay, *Cyanocitta cristata*		8
Eastern bluebird, *Sialia sialis*		8
Mountain bluebird, *Sialia currocoides*		7, 8
Gray catbird, *Dumetella carolinensis*		8
Yellow warbler, *Dendroica petechia*		8
Common yellowthroat, *Geothlypis trichas*		8
Yellow-breasted chat, *Icteria virens*		8
Western tanager, *piranga ludoviciana*		8
Common redpoll, *Carduelis flammea*		8
Porcupine, *Erethizon dorsatum*		8
Mink, *Mustela vison*		3, 8
Domestic or introduced species (N = 10)		
Gray partridge, *Perdix perdix*		7
Ring-necked pheasant, *Phasianus colchicus*		8, 33
Rock dove, *Columbia livia*		2, 6, 8, 10
European starling, *Sturnus vulgaris*		2, 6, 7, 8
House sparrow, *Passer domesticus*		2, 6, 8
Norway rat, *Rattus norvegicus*		8
House mouse, *Mus musculus*		6, 8
Domestic horse, *Equus caballus*		3, 7
Domestic cattle, *Bos Taurus*		3, 7, 33
Domestic sheep, *Ovis aries*		3, 7

* New species associated reported since Kotliar (1999).
a. From Kotliar et al. (1999): PD = prairie dogs as prey or carrion; OP = other vertebrate prey or carrion

found on colonies; BU = burrows for nesting/shelter; OV = open vegetation or bare ground for nesting/foraging; PV = prairie vegetation for nesting/foraging; TP = temporary pools for breeding/forage.

b. Citations: 1 = Koford 1958; 2 = Tyler 1968; 3 = Campbell and Clark 1981; 4 = Clark et al. 1982; 5 = O'Meilia et al. 1982; 6 = Agnew et al. 1986; 7 = Reading et al. 1989; 8 = Sharps and Uresk 1990; 9 = Mellink and Madrigal 1993; 10 = Barko 1996; 11 = Manzano-Fischer 1996 and Manzano-Fischer et al. 1999; 12 = Ceballos et al. 1999; 13 = Kretzer and Cully 2001; 14 = Shipley and Reading In press; 15 = List and MacDonald 2003; 16 = Truett, J. C. 2002; 17 = Restani et al. 2001; 18 = Sidle et al. 2001; 19 = Sheffield and Howery 2001; 20 = Arrowood et al. 2001; 21 = Bak et al. 2001; 22 = Desmond et al. 2000; 23 = Davidson et al. 1999 (Gunnison's prairie dogs); 24 = Price 2002 (white-tailed prairie dogs); 25 = Kolbe et al. 2002; 26 = Ellison Manning and White 2001; 27 = Lomolino and Smith 2003; 28 = Lomolino and Smith 2004; 29 = Cook et al. 2003; 30 = Davis and Theimer 2003; 31 = Cartron et al. 2004; 32 = Holycross and Fawcett 2002; 33 = Smith and Lomolino 2004; 34 = Winter et al. 2003; 35 = VerCauteren et al. 2001; 36 = Desmond 2004; 37 = Dano 1952; 38 = Miller et al. 1996.

7

Economics: How Much Is a Prairie Dog Worth?

The natural world provides us with food, water, timber, stone, medicines, sunsets, mountains, open plains, plants, wildlife, and many other useable resources. We can manufacture none of these, with the exception of some medicines, yet all are intrinsically valuable to us and most are crucial to our existence. From our beginnings, humans have been dependent on the natural world for survival, but the way in which we view this connection is constantly changing.

From earlier days of barter and trade to our current market economy, natural resources have been exchanged by different means; however one thing has remained the same. At the moment when humankind conceived of owning and trading we entered into the dilemma of assigning value. Setting up a system that requires a clear economic value is very convenient for some commodities but not for others. In our modern economy, we assign values to privately-owned commodities and make purchases of those we deem valuable at what we see as a fair price. Most of the time, because we pay for the things we purchase, we have ownership of them and we take care of them. We clean and service our cars to keep them running, we wash our clothes, and repair our houses.

Economic value is relatively easy to assign when ownership is apparent, when someone manufactures something, or when something is rare. But what happens to those commodities and services that are not clearly owned, are never produced or provided by anyone, or are perceived as free or unlimited? How do you assign an economic value to the air when no one ever has to purchase it? What value is assigned to the feeling one gets looking at a mountain stream, a beautiful sunset, or wildflowers? How much is a prairie dog worth?

The preceding questions lead us to a powerful question at the root of our present day dilemma concerning the preservation of species: If something has no assigned monetary value, what is its fate in a world largely governed by the actions of a free-market economy? Answers to all of these questions are becoming more and more crucial to maintaining the integrity of natural systems. Because our modern world speaks largely in numbers, and uses gross national product indicators as the chief metric of societal success, the answer to our root question is, most things without a clear value are overlooked, degraded, or overexploited.

In this chapter, we examine ways to arrive at an economic value for other inhabitants of the natural world, what many people consider priceless. The hope of the environmental economist is that giving value to these forgotten entities enters them into the dialogue of modern society. It does this by including them in the economic system that drives most of our societal decisions and actions.

We discuss the economics of natural entities in three steps. First, we give a short introduction to economics of the environment by considering the economics of species conservation on a general level. Then we apply these theories and perceptions to our focal species, the prairie dog, discussing some economic conflicts concerning prairie dogs and exploring costs and benefits of conserving prairie dogs in light of grazing, recreational shooting, development, and the Endangered Species Act. We also look at economically viable strategies for conserving prairie dogs, such as ecosystem service valuation, watchable wildlife (or ecotourism), education, and relocation programs.

Environmental Economics

Economics deals most explicitly with monetary value: income and expenditure, how much humans are willing to pay, and what costs they are willing to accrue to conserve species. However, there are other types of values that are harder to address at a societal level because we have no currency specifically devoted to them. These are moral and ethical values that stem from a wide variety of cultural, educational, philosophical, and personal beliefs and experiences. To some extent the economist believes these are reflected in what people choose to do with their money, however, it is important to know that there does remain a distinction.

The challenge of determining value for a species is captured in the questions asked by Harold Morowitz (1991), who asked how much a species is worth and what is the value of a species to human society? He pointed out that we often are left with balancing what is ethically good with what is economically good. This presents the most confounding problem with the conservation of species that faces us today, the problem of connecting the "good of ethics with the goods of economics" (Morowitz 1991). Ethics inherently involves thinking about the public good and the good of future generations while economics is traditionally the exchange of private goods over short periods of time. In modern society, our actions are largely governed by the free-market economy where private commodities are exchanged and value is freely altered based on supply and demand. However, the natural world operates on temporal and geographic scales much larger than commonly considered in our traditional economic system.

In addition, natural resources like air, water, and biological diversity are public rather than private goods. They may not be owned by any single entity and so are not easily incorporated into private markets. For example, the cost of degrading our natural heritage is not usually factored into the process of producing private goods. A paper mill traditionally does not include diminished public air quality in its production costs (unless through governmental regulations). No one clearly owns the air that is being polluted and, even if it was clear whom to pay, it is difficult to estimate the monetary value of diminished air quality.

This is true for most of our natural resources, including biodiversity and habitat for species. The result is that we usually fail to assign value to non-human species and natural services, which allows producers to degrade the natural world at no cost. This in turn promotes unsustainable practices by lowering the cost of production and thus artificially elevating the rate of production (Figure 7.1a, b).

The term *externality* is used to define the costs of degradation to public goods that are not easily quantified, occur external to the site of production, and are therefore left out of the economic functions of the supply and demand curves (C in Figure 7.1b). Some examples of externalities are the cost to the public of air and water pollution from industrial run-off, the cost to the public of habitat that is destroyed for the development of a shopping center,

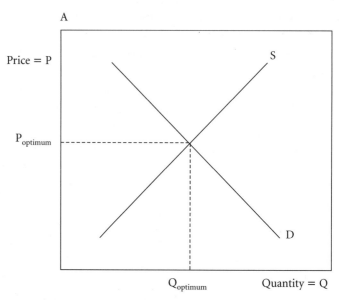

Figure 7.1a. A normal supply and demand graph. S = supply curve, D = demand curve, Y-axis is currency (i.e., price), X-axis is quantity of a good. As supply decreases for a certain product, demand increases or vice versa. The point where supply of a good is in equilibrium with demand is the optimum quantity and price ($Q_{optimum}$, $P_{optimum}$). As supply exceeds demand (moving to the right of the optimum) there is waste, or extra cost to the supplier. If supply does not meet demand (moving to the left of the optimum) buyers pay more than the optimum for the product.

and the cost to future generations of a species that goes extinct due to the building of a dam. If producers are allowed to count the cost of externalities as zero, consumers will buy more of the produced goods or services because they will be less expensive than they "should" be. This artificially elevates the demand curve, and producers will be able to use the money they would have paid for the externality to produce more goods, which further degrades the natural resource (Mills and Graves 1986)(Figure 7.1b). This is the fundamental economic process that encourages unsustainable destruction of habitats, and species, and ultimately leads to the rapid decline in biodiversity we are currently experiencing.

Since the 1990s, the field of environmental economics has grown largely in response to increased public awareness of the biodiversity crisis and the increased ethical values placed on conserving rather than exploiting natural

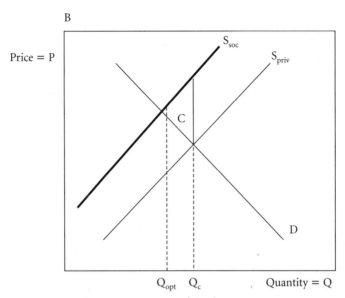

Figure 7.1b. A supply and demand graph with a supply curve that represents social costs of producing a product (S_{soc}). S_{priv} = supply curve of the private producer that does not pay the full social costs of producing his product. In this scenario the cost of supply is artificially low because the producer does not pay the full cost to produce his goods. This causes artificially high production and thus lower prices to the consumer (Q_c). Q_{opt} = Q optimum, the equilibrium price with the social cost of supply factored in. C = cost to society from overproduction of a good at an artificially low cost when equilibrium falls at Q_c.

resources. Today environmental economists tackle the challenge of incorporating preservation of the natural world into the mainstream market. This is done in two major ways: Determining the value of externalities is a free-market tactic that seeks to curb degradation of natural resources by "correcting" the costs of production (by accounting for the externalities) through pollution taxes or credits that readjust the supply and demand curves (i.e., quantifying C in Figure 7.1b to shift the supply curve from S_{priv} to S_{soc}). Another way of curbing environmental degradation is more policy oriented and relies on the institution of government regulations like the Clean Air and Water Act (Stavins 1991).

Whether policy-oriented or free-market tactics are employed, the key issue is still determining the monetary "value" of the externality. This brings us to the following key questions: How do we place a value on these natural elements? And how do we effectively incorporate both ethical good and economical goods into our determination of value?

There are many ways in which humans value the natural world. In the most of simple terms, natural entities can have use and non-use values (Williams and Diebel 1996). Use values directly affect the resource in question and can be consumptive or non-consumptive. For example, use values of prairie dog towns could be recreational shooting, which causes a reduction in the resource and is consumptive; or eco-tourism, a recreational activity that provides direct benefit to humans but does not consume the resource (Table 7.1). Non-use values are less tangible and could be the maintenance of ecological processes in the prairie ecosystem by prairie dogs, the benefit of having prairie dogs left for future generations to witness, or the beauty of flowers blooming on prairie dog towns (Table 7.1).

Among their many writings on the issue, Paul Ehrlich and E. O. Wilson present three compelling economic arguments for conserving species—some are use values and some are non-use values (Ehrlich and Wilson 1991). The first argument is largely ethical stating that humans have a moral responsibility to "protect our only known living companions in the universe." Human

Table 7.1.   Use and non-use values of prairie dog towns.

Use values

Consumptive	Non-consumptive
Recreational shooting	Watchable wildlife
Grazing	Hiking
Open land with development potential	Wildlife photography
Plant/herb collecting	Research
Food: prairie dogs and associated species	Education

Non-use values and natural services

Cultural symbol of the Great American prairie
Opportunity for future generations to witness
Biodiversity
Nutrient cycling
Soil aeration
Water filtration
Prey base
Carbon sequestration
Existence

beings have a natural appreciation for the beauty of nature and attain fulfill-ment from being around and among the natural inhabitants of this world. The authors cite the popularity of pet keeping, gardening, and wildlife watch-ing as proof of most humans' innate desire for contact with other species. They also make the point that these activities generate a substantial amount of economic productivity.

To the above argument, Morowitz (1991) adds a class of species that has significant value to humans because of its cultural value or its ability to inspire humankind. This is the class of species that people readily open their pocketbooks to conserve the so-called "charismatic megafauna," such as the large African mammals, the American bald eagle, the humpback whale, and the wolf. Studies examining willingness to pay for conservation (these studies ask people how much they would pay to protect a species, for example) show us that different species can have economic value by being a cultural symbol or a symbol of wilderness. Contingent valuation is another technique used to arrive at non-consumptive value of a species (Portney 1994). Contingent valuation uses general public surveys to gather information about how much people are willing to pay for hypothetical projects or programs. These questionnaires or surveys can be adapted to species conservation programs. However, a common problem with these surveys is that people tend to say they will pay more in a hypothetical situation than they are actually willing to pay in a real situation.

While we cannot precisely quantify moral values and ethical mandates, we can indirectly arrive at a value for different species by measuring how much people pay to save or to see different species; we can assess how much money is paid to conservation organizations or for ecotourism. On a governmental scale, we can assess moral value through quantifying how much money is used to enforce regulations and protection acts like the Migratory Bird Treaty Act and the Endangered Species Act.

The second and third reasons mentioned by Ehrlich and Wilson for conserving species are more economically based and stem from consumptive rather than non-consumptive uses (Ehrlich and Wilson 1991). The immense number of species that have evolved over millions and millions of years are a non-renewable resource that is still largely untapped by humankind. We have cultivated our staple foods wheat, rice, and corn from this warehouse of biodiversity. We have also found medicines for all kinds of ailments from

malaria to cancer. Along with the demise of species is the demise of oppor-
tunities for products that we could market. The argument on the grounds
of potential future sources of natural products is the most widely used and
directly applicable economic argument for species preservation (Eisner
1991). By comparing costs of biological harvest with the cost of chemical
research, or biotechnology efforts needed to produce a similar substance, we
can directly arrive at monetary values.

An important new way of deriving value of natural elements and services is
through the concept of ecosystem services (Daily 1997; Sala and Paruelo 1997).
While humanity has been intuitively aware for centuries of the dependence of
human societies on pest control, pollination, climate regulation, soil retention,
flood control, soil formation, and other important and life-sustaining services
that earth's non-human life forms and systems provide; ecosystem services,
as a technical, cohesive form of study, was formed during the 1970s (Mooney
and Ehrlich 1997). Ecosystems form the life-support systems for our earth and
the basic elements that humans need for survival. They maintain and generate soils, clean and recycle water, naturally control pest species that can attack
crops, and much, much more. However, these services, and the extent to which
humans compromise these services through conversion and degradation of
natural lands, are not accounted for in the economic market.

The concept of ecosystem services has contributed to a growing body of
literature examining the value of species and ecosystems through asking the
question, What do we lose in terms of free natural services when we convert
natural lands to more human landscapes like agriculture, grazing lands, or
cities and towns? For example, Costanza et al. (1997) used published studies
on ecosystem services to estimate the value of 16 different ecosystem services
in 17 different biomes. The total worth of these services, most of which fall
outside the global economic market, they estimated as $33 trillion per year
(U.S. dollars), which they compare to the average Global Gross National
Product (GGNP), estimated as $18 trillion per year. Costanza et al. (1997)
performed the first, and so far the only, widespread determination of ecosystem service values in the 17 biomes and show that globally, services provided
by natural environments are almost double the value of the entire global
economic output. They valued global grassland biomes at $232 per hectare/
year (Costanza et al. 1997). Goulder and Kennedy (1997) offer other examples
along with background theory. Thus, according to these preliminary estimates,

human societies are receiving and freely degrading natural services that are almost twice as valuable than all present economic activities combined.

Of all the above described valuation techniques, research on ecosystem services of prairie dogs and their role in maintaining native grasslands holds the most promise for incorporating prairie dogs into our current economic system because prairie dogs are such major players in maintaining native grassland functions. Because of this, we provide a more in-depth discussion of ecosystem services provided by native grasslands in the following sections of this chapter. We also discuss the role of prairie dogs in maintaining and promoting these services and ways of using ecosystem service valuation to encourage landowners to conserve prairie dogs rather than eradicate them.

Prairie Dog Economics: Grazing

One of the most pivotal controversies surrounding the conservation economics of prairie dogs is the issue of costs to livestock owners. In 1902, C. H. Merriam surmised that prairie dogs ate between 50% and 75% of the forage that could be used by livestock (Merriam 1902). Following directly on the heels of this announcement was a strong campaign in the United States to eradicate prairie dogs from grazing allotments on public lands. This campaign was largely subsidized by the U.S. government starting in 1915 with money given to the Biological Survey to poison prairie dogs and culminating in the formation of the Predatory Animal and Rodent Control division in 1929. In 1931, the Animal Damage Control Act was passed, which sanctioned both poisoning of prairie dogs and cooperation between the government and private parties in poisoning campaigns (Miller et al. 1994). Authority for poisoning prairie dogs transferred to United States Fish and Wildlife Service and then finally to the Department of Agriculture in later years.

The costs to taxpayers for these poisoning campaigns are not trivial. Typically, livestock grazing permittees pay only 5%–10% of the poisoning cost, leaving 90%–95% of the cost to public funds, a huge externality cost borne by the public (Roemer and Forrest 1996). Because poisoning is over-seen by a wide range of federal, state, and private entities that differ between states, it is difficult to arrive at an overall estimate of the amount taxpayers have spent on poisoning efforts over the past 100 years. However, several examples of federal and state spending on poisoning are available.

Costs for prairie dog poisoning are wide-ranging depending on technique and cost of labor involved. Costs have been computed to range from $11.05–$75 per hectare depending on the type of poison used (Andelt 2006). Generally, grain-fed poisons are cheaper than fumigants. From 1980 to 1984, $6,200,000 was spent by the U.S. government to poison 185,600 hectares of prairie dog habitat in South Dakota alone (Miller et al. 1994). This was the most expensive federally subsidized poisoning campaign at about $3 per prairie dog and $33.41 per hectare (Sharps 1988). The Nebraska National Forest poisoned prairie dogs on 55,756 hectares from 1978 to 1992 at a cost of $616,000 or $11.05 per hectare (Roemer and Forrest 1996). A total of 13 million hectares of prairie dogs was poisoned under federally supervised programs between the years 1916 and 1920 (Bell 1921). Using the figure from the Nebraska National Forest as an estimate of average government spending, today's equivalent of at least $143,650,000 of taxpayers' money was spent on poisoning prairie dogs between 1916 and 1920.

As late as 1991, Congress appropriated another $256,000 for poisoning prairie dog towns on the Cheyenne River and Rosebud Indian reservations in South Dakota (USBIA 1991). Overall, government spending for poison control programs between 1978 and 1992, for the states of Montana, South Dakota, and Wyoming combined, totaled well over $10 million (Roemer and Forrest 1996). This estimate only includes federally conducted programs in these states, and does not include the many government assistance programs available for private landowners, including state, county, and municipality expenditures and incentives.

Estimates of state government spending are lower in total than federal spending but tend to be higher per hectare. For example, state programs in the Great Plains states of Montana, South Dakota, and Wyoming expend $36,500–$50,000 annually on poisoning programs at a cost of between $21.79 and $25.77 per hectare. In addition, state governments in some states, such as South Dakota and Wyoming, pass expenses off to private landowners with ordinances requiring private landowners to poison, at their own expense, "prairie dog infestations" that occur on their lands. Fines for private landowners that do not comply with poisoning efforts range from $50.00 per day to $2,500 per day (Roemer and Forrest 1996).

New research conducted in the early 1980s and 1990s significantly counteracts both the need for poisoning campaigns and the economic profitability of conducting these campaigns. Several studies show that the presence of prairie dogs may benefit livestock through increasing forage digestibility, nutrients, and the abundance of forage preferable to livestock. Prairie dogs also control mesquite and prickly pear cactus, plants that reduce the availability of grass for cattle (O'Meilia et al. 1982; Coppock et al. 1983; Kreuger 1986; Bonham and Lerwick 1976). The clinching argument against poisoning campaigns is one that counteracts C. H. Merriam's original estimate of forage competition between livestock and prairie dogs. A study in 1985 found only a 4%–7% level of competition between livestock and prairie dogs, an order of magnitude less than Merriam's original estimate (Uresk 1985). This study is supported by the conclusions of two previous studies that showed no significant difference in weight gain of steers that graze on prairie dog towns as opposed to off prairie dog towns (Hansen and Gold 1977; O'Meilia et al. 1982). A study done in 2006 showed that in pastures that had 20% of their area occupied by prairie dogs, the estimated value of weight gain of cattle was reduced by $14.95 per steer (from $273.18 to $258.23) and in pastures that had 60% of their area occupied by prairie dogs, the estimated value of weight gain of cattle was reduced by $37.91 per steer, for a total reduction of 5% at the lower level and 14% at the higher level of occupancy of prairie dogs (Derner et al. 2006).

Prairie dogs do not seem to significantly threaten the profits of livestock owners, but costs to the public in the form of externalities are in the millions of dollars for prairie dog poisoning efforts. Thus it does not make economic sense to continue poisoning prairie dogs. In fact, research suggests that poisoning campaigns operate at an economic loss. For example, a more detailed study on the economics of prairie dog poisoning conducted in South Dakota concludes that poisoning is an unsustainable economic practice both from a rancher's perspective and from the perspective of the United States Forest Service. An economic analysis of prairie dog control in the Conata Basin of South Dakota considered the ranchers' point of view by balancing the costs of a poisoning campaign with the benefits of increased forage in Animal Unit Months (AUMs). The analysis also considered that the United States Forest Service, representing the public, had other costs as well as poisoning, including environmental impact statements for other animals that would be affected by poisoning, as well as monitoring costs. The results of the economic analysis, conducted over five consecutive years, showed that control costs would not be recovered for 40 years for the rancher and 22 years for the United States Forest Service, if there was an annual repopulation rate of prairie dogs that was 5% or more. A more realistic expectation for prairie dog recovery rates is about 30%. At prairie dog recovery rates of 10% or more the analysis showed that the cost for poisoning could never be recovered by either the rancher or by the United States Forest Service (Collins et al. 1984).

Government subsidized poisoning campaigns are a good example of an instance when externalities are not accounted for (Figure 7.1b). Grassland permittees, who comprise fewer than 2% of the American public, are only required to pay 5%–10% of the total cost of eradication rather than 100%. Even at this subsidized rate, it takes at least 40 years for a rancher to pay off the cost of poisoning at low repopulation rates, and at normal prairie dog repopulation rates the benefits of poisoning will never exceed the costs (Roemer and Forrest 1996).

Some alternatives to poisoning prairie dogs have been suggested, although no cost/benefit analyses of these alternatives have been conducted. In light of evidence showing seemingly little or no significant competition between livestock and prairie dogs, one alternative suggests that federal funds currently being used for poisoning campaigns be used instead to subsidize conservation programs (Miller et al. 1990). These programs would be designed to encourage ranchers to integrate prairie dogs into their grazing management regimes instead of eliminating them. Benefits of allocating funds to this purpose include healthier grassland ecosystems and increased protection for listed species on prairie dog towns.

In those instances where there is no alternative but to control prairie dogs, non-lethal methods have been suggested. Placing visual barriers around areas where prairie dogs are not desired is emerging as a possibly effective way to control direction and speed of expansion of prairie dog towns. Both natural barriers, such as dense vegetation, and artificial barriers, such as burlap fences are highly effective (with an effectiveness of 60%–80%) at deterring expansion of prairie dogs (Franklin and Garrett 1989). However, it is unclear how successful these efforts would be in the long-term. More research on this is needed, including the required width of tall grass needed to prevent expansion and the use of contraceptives (Andelt 2006).

Prairie Dog Economics: Recreational Shooting

Recreational shooting is another issue that needs to be considered when analyzing the costs and benefits of prairie dog conservation. Recreational shooting of prairie dogs can bring in a substantial amount of revenue to towns that promote this business. For example, in South Dakota shooters spent an average of 46,000 hunter days shooting prairie dogs per year. The average shooter spent $70 a day for a total of $3.2 million in general revenue to towns and businesses (Sharps 1988).

Many land management agencies responsible for North American grasslands have mandates to provide recreation while ensuring good forage for permitted grazers. Thus, recreational shooting has been permitted on many multiple-use lands. However, research indicates that recreational shooting of prairie dogs can have extremely detrimental effects on prairie dog population levels, stress levels, and colony viability (Knowles 1982; Knowles 1988; Reeve and Vosburgh 2006). In a study conducted in Montana, population size of colonies decreased by 35% on hunted colonies and 15% on adjacent unhunted colonies (Vosburgh and Irby 1998). Given the declining numbers of prairie dogs, and ethical and moral issues associated with this practice, it is difficult to justify shooting of prairie dogs in the name of recreation.

Prairie Dog Economics: Other Species

Perhaps the biggest externality that is not accounted for when considering the destruction of prairie dogs and their habitat is the potentially immense deleterious effect on dozens of different species associated with the prairie dog habitat. As mentioned in chapter 6, prairie dogs have an important effect on species, and thus prairie dog towns are crucial habitats on which many other species depend. The potential loss of this great source of biodiversity is a serious and expensive scenario in ecological, ethical, and economic terms.

There is no widely comparable and obtainable information available on the money being spent by state agencies, and non-profit environmental groups on conservation of non-listed species that are dependent on prairie dog towns for survival. Information on the eight non-listed species Kotliar et al. (1999) propose as linked to prairie dog decline (burrowing owl, mountain plover, ferruginous hawk, golden eagle, swift fox, horned lark, deer mouse, and grasshopper mouse) is needed to more fully estimate the costs of prairie dog decline on biodiversity.

There is quantifiable information on the costs of conserving species under the Endangered Species Act (ESA). Costs for ESA listing to the federal government include listing costs, costs of recovery plans, and costs of habitat acquisition. Costs for paperwork and legal work to list species under the endangered species act average $107,000 per species with total costs for all species currently listed at $2.26 billion (Burnett and Allen 1998). Opportunity costs of listing greatly vary and have not been quantified for most species (opportunity costs are the costs incurred by foregoing the next best alternative). However, opportunity cost estimates to the public for conserving the northern spotted owl ranged from $450 million to $46 billion depending on the community affected. Opportunity costs included value of foregone timber and the associated wood product producing industries and loss of jobs in the timber industry and associated industries (Shogren 2000). State and federal costs for endangered species recovery from 1989 to 1996 totaled over $5.6 billion ($.8 billion per year), not including land acquisition costs and by fiscal year 2004 the annual total was $1.35 billion for all species with annual land acquisition costs totaling $136 million (USFWS 2004). Individual recovery plan costs range from $145,000 to $700 million depending on the species (USFWS 2004).

Currently, there is one federally threatened or endangered species, the black-footed ferret *(Mustela nigripes)* that depends on prairie dog towns. Its demise is directly linked to the fall of prairie dog populations over the last 100 years and it cannot continue to exist without prairie dogs. However, federal and state endangered species expenditures reports show that spending on black-footed ferret recovery greatly exceeded spending on Utah prairie dog recovery even though prairie dog habitat is essential to black-footed ferret survival (Tables 7.2 and 7.3) (USFWS 2004). In addition, black-footed ferret survival likely depends more on the survival of large Gunnison's and black-tailed prairie dog complexes. which receive no money through the Endangered Species Act. In total, ESA expenditures reported in 2004 for black-footed ferret and Utah prairie dog recovery alone were $4.1 million, which is less than 1% of expenditures on all 1,260 species expenditures reported in that year (Table 7.2) (USFWS 2004).

In addition to species that are already endangered on prairie dog towns, there are at least two more species, burrowing owls and ferruginous hawks, which are rapidly declining and are intricately associated with prairie dog towns. This sets up an interesting economic and ecological dilemma. A recovery plan that focused on saving prairie dog habitat rather than each individual species whose decline is associated with prairie dogs could effectively reduce six potential recovery plans (recovery of the black-footed ferret, swift fox, burrowing owl, mountain plover, ferruginous hawk, Utah prairie dog) to one, saving tax payers between $725,000 and $3.5 billion (cost of individual recovery plans range from $145,000–$700 million).

Prairie Dog Economics: Prairie Dog–Friendly Business

One popular and economically viable alternative to eradicating prairie dogs is to promote ecotourism or watchable wildlife sites. Many cities including Flagstaff, Arizona, and Santa Fe, New Mexico, have successfully integrated prairie

Table 7.2. Endangered Species Act expenditures (FY 2004) for the black-footed ferret *(Mustela nigripes)* in-situ and experimental nonessential reintroduced populations (EXPN), and Utah prairie dog *(Cynomys parvidens)* in real U.S. dollars.

Species	Type	Cost (U.S. $)	Cost rank*	Cost percentile*
Mustela nigripes	State total	2,800		
Mustela nigripes (EXPN)	State total	302,950		
Mustela nigripes	Federal total	1,059,513	100	10
Mustela nigripes (EXPN)	Federal total	1,716,626	77	10
Mustela nigripes	Species total	1,062,313		
Mustela nigripes (EXPN)	Species total	2,019,576		
Cynomys parvidens	State total	87,800		
Cynomys parvidens	Federal total	942,254		
Cynomys parvidens	Species total	1,030,054	102	10
Cynomys parvidens	Federal land acquisition	1,000	104	10
Both species	Grand total	4,111,943		
All listed species	State total	60,965,476		
All listed species	Federal total	732,154,431		
All listed species	Species total	793,119,907		
C. parvidens + *M. nigripes*	% Total ESA expenditures: .5%			

* Rank out of 1,260 species expenditures. Both fall within the top 10% of species for costs.

dog towns that exist within the city limits by designing interpretive trails for the public to enjoy. In addition, Montana has a state monument on Interstate 91 that is a very popular stopping place for tourists to watch prairie dogs. Prairie dogs are excellent wildlife subjects to watch because they are social and diurnal (out during the day). Prairie dog watching is extremely popular in some states and thus has the opportunity to be exploited for economic gain. However, urban colonies may not provide good homes for many associated species that do less well near people, meaning that while watchable wildlife sites could save prairie dogs, they won't necessarily save large complexes but will likely only save remnants that do little for the survival of the species as a whole, and the survival of associated species.

Open space ordinances could also be targeted to conform with prairie dog colony boundaries, thus preserving open space near developments and conserving prairie dog towns. It is not clear whether retaining open space

Table 7.3.   Reporting Government agency expenditure on *C. parvidens* and *M. nigripes* for fiscal year 2004.

Agency	Total expenditure	Rank
Dept. of Agriculture	206,914	3
Dept. of Commerce	0	10
Dept. of Defense	5,500	5
Dept. of Energy	1,200	7
Dept. of Homeland Security	0	10
Dept. of Interior	2,126,328	1
Dept. of Transportation	565,400	2
EPA	400	9
Federal Communications Commission	0	10
Federal Energy Regulatory Commission	700	8
Nuclear Regulatory Commission	2,100	6
Smithsonian	197,900	4
Tennessee Valley Authority	0	10

Source: U.S. Fish and Wildlife Service Federal and State Threatened and Endangered Species Expenditures Fiscal Year 2004.

Notes: All amounts are actual dollar amounts judged by the reporting agency to be reasonably attributable solely to that species, and include land acquisition costs. Thus, expenditures that include *C. parvidens* and/or *M. nigripes* in conjunction with other species are not included but are reported as a lump sum for all species under "Other ESA" expenditures in the above cited report. Rank equals the order of total expenditure from most (1) to least (10) for each reporting agency. This shows which agencies are dealing with prairie dog conservation issues in the greatest amounts.

in lieu of developing an area can actually be more profitable than a development would be. However, the extended benefit of open space is often cited as an important factor that contributes to quality of life. Conner et al. (2002) cite studies that quantify home values adjacent to open grassland in the American West as 7%–34% higher than comparable houses that are not adjacent to open space. In addition, large units of private rangelands cost county and state governments less per dollar of revenue provided than if the rangeland is divided into smaller ranchettes (Conner et al. 2002). Thus, offering incentives for conservation-friendly ranchers and farmers to keep large blocks of land from development could help prairie dogs and government bottom lines.

Another alternative to prairie dog eradication is relocation of prairie dog towns in conflict areas to areas that have a lower level of conflict with humans. A model city is Boulder, Colorado, which in addition to conducting prairie dog relocations has developed a management plan for deciding what to do with prairie dog towns within and near the city. Costs of prairie dog relocations vary depending on the size of the town to be relocated as well as the method of relocation used.

The two most popular methods of prairie dog relocation are soaping and trapping. The average cost of soaping a prairie dog town depends on water costs and size of the town. Trapping is more expensive then soaping. Traps are about $35 each (in 2004) and labor for trapping is equivalent to about 300 human-hours per 100 prairie dogs trapped. Preparation costs for the relocation site may vary as well. If relocating to a former prairie dog town with intact burrows, site preparation is zero. Vacant fields without burrows however require labor and machinery to auger artificial burrow systems, install and bury nest boxes and plastic pipes to enable prairie dogs to adjust to the site and eventually dig natural burrows. Success rates for translocation to fields with no holes or burrows increase if prairie dogs are penned on the colony for a few days.

Volunteers have been successfully used in most prairie dog relocation efforts. Volunteers provide an inexpensive source of labor. In addition, volunteers become more invested in prairie dog issues through this process and become more informed citizens capable of making more educated decisions in future prairie dog conflicts. The benefit of this education is enormous in terms of more efficient and well-planned future county, city, and town management.

Prairie Dog Economics: Ecosystem Services of Native Grasslands

While relocation, open space ordinances, and prairie dog-watching businesses described earlier are helpful in maintaining prairie dog populations
they are a bit like applying bandages to a severed limb. There are limited
places to relocate prairie dogs, especially as land conversion increases through
time. In addition, watchable wildlife sites, while they keep prairie dog towns
in existence, can be small and essentially non-existent in terms of functioning as parts of larger biological populations that contribute to the ecological
integrity of native grasslands.

Effectively ensuring the continued existence of prairie dogs at population
and metapopulation levels will require conserving large expanses of native
grasslands. This means directly curtailing the conversion of these habitats
to other uses, and reclaiming some of these habitats back to grassland from
ranchland, farmland, urban, or other uses. Native grasslands provide many
natural benefits including increased water recharge to aquifers, soil mixing
and building, nutrient cycling, productivity, carbon and methane sequestration, and climate control (Table 7.4). It is not yet clear how much, or
which, grassland services depend on prairie dogs for continuance (Table 7.4).
However, an exploration of the magnitude of benefits provided by natural
grassland services makes one think very seriously about the risks to human
society of promoting continual decline of prairie dogs and their habitat.

Exploring prairie dog economics at the larger scale of grassland ecosystem
services provides a better estimate of prairie dog values than comparisons
previously discussed in this chapter (grazing, recreational shooting, open
space, ecotourism). It does so because it places prairie dogs within the context
of the larger ecological system, of which they are strong contributors. In addition, exploring and quantifying ecosystem services, and the role that prairie
dogs play in the delivery of these services provides an avenue in which to
structure economic incentives for conserving prairie dogs. By examining what
is lost when we convert native grasslands to agriculture, ranchland, or urban
and rural environments, we can arrive at values and compensatory mechanisms for promoting the value of unaltered lands. For example, this could
lead to offering payments or tax breaks to landowners, based on valuation of
the natural services that their lands provide, and other economic incentives
to keep the land in natural condition and prairie dog towns intact.

Grasslands are historically the most prevalent and dominant biome in the United States, comprising 40%–50% of the contiguous land base (Conner et al. 2002). Humans derive an enormous number of goods and services from American grasslands. Products that are accounted for in the Gross National Product are derived mostly from ranching and agricultural activities. The most prevalent goods produced in grasslands, or former grasslands are agricultural crops, meat, milk, leather, and wool. However, the value of natural grasslands in terms of life-supporting services is immense, sometimes more than the total amount of products that have market values. Yet, these services are largely unaccounted for in the standard market (Sala and Paruelo 1997).

Quantifying the costs associated with conversion of natural or lightly grazed grasslands to agriculture, heavily grazed lands, or urban and rural development clearly illustrates the magnitude of benefits derived from natural grassland services that are presently unaccounted for in human society. The major services provided by grasslands are maintenance of a healthy atmosphere through sequestration of gases in the soil, water infiltration and supply, erosion control, soil formation, waste treatment (through decomposition), pollination, biocontrol, food for humans and other species, maintenance of biodiversity, recreation, and weather amelioration (Costanza et al.1997; Sala and Paruelo 1997; Conner et al. 2002) (Table 7.4).

Values of ecosystem services are determined by examining the consequences of losing the service and then estimating the value of that loss. For example, grasslands help to maintain the proper atmospheric composition necessary for earth's current species to survive. They do this by sequestering large amounts of the gases carbon, nitrous oxide, and methane as organic matter in the soil. These gases are by-products of production, both human and non-human, and their rise in today's atmosphere is contributing to the current global warming trend that threatens to increase severe weather patterns, interfere with pollination, migration, breeding, and blooming cycles, disrupt water cycles, and cause severe and catastrophic economic loss on a global scale. Cultivating natural grasslands, converting these grasslands to residential development, or heavily grazing grasslands can release these carbon stores and lower soil storage capacity by tilling, compacting soils, and lowering productivity levels of vegetation (Sala and Paruelo 1997; Conner et al. 2002).

Between 20%–50% of organic carbon was lost (released) from soils in the first 40–50 years of cultivation in the Great Plains. While reconverting cropland

Table 7.4. Ecosystem services provided by natural or lightly grazed grasslands.

Service	Value/ha	Notes	References	Prairie dog influence
Carbon sequestration	$7–$200		$7: Costanza et al. 1997; $200: Sala and Paruelo 1997	Likely
Nitrous oxide sequestration	$28.50		Sala and Paruelo 1997	Likely
Methane sequestration	$2.70		Sala and Paruelo 1997	Likely
Water infiltration and precipitation	$3.00	Infiltration could be higher on prairie dog towns due to burrowing activities that aerate the soil. Conner et al. 2002 state that over 50% of U.S. cities rely on groundwater.	Costanza et al. 1997; chap. 6 in this book	Definitely
Erosion	$29–$100		$29: Costanza et al. 1997; $100: Sala and Paruelo 1997	Likely
Soil formation	$1.00	Could be higher on prairie dog towns due to increased nutrient cycling, primary productivity, soil aeration, and increased soil productivity.	Costanza et al. 1997; chap. 6 in this book	Likely
Waste treatment	$87.00		Costanza et al. 1998	Unknown
Pollination	$25.00		Costanza et al. 1999	Likely
Biocontrol	$23.00		Costanza et al. 2000	Unknown

	Value/ha	Notes	Source	Certainty
Food	$67.00		Costanza et al. 2001	Unknown
Recreation	$2.00	Global figure. Likely higher in U.S. grasslands. Conner et al. 2002 gives a figure of $37 million/year generated from recreation in grassland states.	Costanza et al. 2002	Likely
Biodiversity	Not available	9 species dependent on prairie dogs and declining. In FY 2004, Endangered Species Act spending on black-footed ferret and Utah prairie dog was $4.1 million.	Kotliar et al. 1999; Table 7.2 in this chapter	Definitely
Weather amelioration	Not available	In U.S. Central Plains study, converting grasslands increased temperature and precipitation. In study of the Sonoran grasslands on the U.S.-Mexican border, higher temperature, lower precipitation, higher desertification/erosion in Mexico were due to more grazing.	Sala and Paruelo 1997	Likely
Forage productivity	Not available	Increases on prairie dog towns	See chap. 6 in this book	Definitely
Nutrient cycling	Not available	Increases on prairie dog towns	See chap. 6 in this book	Definitely
Total value/ha	**$275.20–$539.20**			

back to grassland can increase carbon sequestration again, the degree to which soils recover is variable. Recovery depends on a variety of abiotic and human-use factors, and takes more than 50 years, with some grasslands never recovering their full potential (Sala and Paruelo 1997).

Carbon loss from conversion of grassland to cropland since the mid-1800s is estimated to have contributed one and a half times more CO_2 to the atmosphere than all fossil fuel burning through 1950. Through estimating the costs of the negative effects that increasing CO_2 has on the climate, various authors have estimated the value of carbon sequestration in grassland systems to range between \$7–\$200 per hectare (Costanza et al.1997; Sala and Paruelo 1997; Conner et al. 2002) (Table 7.4). This is significant enough that international discussions on climate change have resulted in proposals to offer carbon credits, or subsidies, to farmers and ranchers who maximize carbon sequestration on their lands (SWCS 2000).

Currently "perverse" subsidies (Myers and Kent 1998), meaning subsidies that cause adverse affects on the economy and the environment in the long-run, encourage farmers and ranchers to convert grasslands even when the output from these lands is only marginal. Subsidized poisoning of prairie dogs, discussed earlier, is one example of a perverse subsidy. Another example is government subsidies to farmers that encourage them to convert parts of prairie dog grassland habitats to agriculture lands in very arid southwestern states where the cost of irrigation, nutrient enrichment, and de-salinization is very high. Replacing these "perverse" subsidies with subsidies that encourage the retention of ecosystem services is a potential economic mechanism that would help to curb global warming, and indirectly result in the conservation of prairie dog habitat through retention of native grassland habitats.

Using estimates of value in U.S. dollars for services in grasslands (Table 7.4) we estimate total value of \$275–\$539 per hectare. This is similar to Costanza et al.'s (1997) estimate of \$232 per hectare and is a conservative estimate because not all known services in grasslands have been quanitified. We also computed rough estimates of the market value of livestock and crops for all states that contain prairie dog habitat and used estimates of total cropland in these states to assign a dollar value per hectare for crops in these states (Table 7.5). No information was available on total public and private livestock range areas by state so we calculated these figures on a per hectare basis using estimates of land cover and livestock total market value on a national scale (Table 7.5).

We estimate average total market value of crops at $471 per hectare for states within the prairie dog range, and for livestock the total market value for the United States we estimate at $430 per hectare. These figures likely over-estimate the value per hectare as they include profits from farms that were not originally on grasslands, feedlots, and high-density farms such as poultry operations that have a low presence in grasslands yet have a high value per hectare. In addition, these estimates of market value are not corrected for subsidies. When costs of production are factored in, profit declines substan-tially. For example, Conner et al. (2002) estimate that when all costs (i.e. production, overhead, opportunity costs of the land) are considered, ranchers in grassland states operate at a deficit (loss) of $351.98–$667.36 per bred cow. This estimate does not factor in the cost of lost ecosystem services on these lands due to conversion or degradation of grassland, a cost of $232–$539 per

Table 7.5. Market value of crops and livestock in states in prairie dog species range.

State	Total cropland (ha)*	MV crops ($1,000)*	MV livestock ($1,000)*	MV crop /ha
Arizona	509,805	$1,587,775	$807,672	$3,114.47
Colorado	4,658,403	$1,216,278	$3,308,918	$261.09
Kansas	11,934,977	$2,418,447	$6,327,797	$202.64
Montana	7,399,468	$733,324	$1,148,791	$99.10
Nebraska	9,098,433	$3,388,265	$6,315,392	$372.40
New Mexico	1,040,343	$397,257	$1,302,773	$381.85
North Dakota	10,708,617	$2,460,372	$772,994	$229.76
Oklahoma	5,996,716	$819,078	$3,637,326	$136.59
South Dakota	8,208,487	$1,575,910	$2,258,715	$191.99
Texas	15,617,715	$3,731,751	$10,402,993	$238.94
Utah	835,245	$257,797	$858,101	$308.65
Wyoming	1,207,881	$137,776	$726,111	$114.06
Total	**77,216,089**	**$18,724,030**	**$37,867,583**	
			Average $/ha = $470.96	
			MV livestock/ha: $429.80**	

Source: 1997 Census of Agriculture, U.S. Dept. Agriculture, USDA land cover estimates, USA 1997.

*2002 Census of Agriculture, Department of Agriculture. MV = market value

**Derived by dividing MV livestock U.S. total/ha pastureland U.S. total. Ranchland areas not available by state.

hectare that farmers and ranchers are not required to pay but that will likely increase farming costs for future generations.

The above information makes it clear that, even with subsidies, ranchers are in an unstable position, and often operating at a loss. However, if ranchers and farmers are forced to subdivide un-profitable lands and sell them for residential development, the resulting fragmentation and conversion of habitat could cause greater and more permanent losses of prairie dogs and associated biodiversity because it would mean loss of large prairie dog complexes, key places for ecological processes on grasslands. An alternative is to encourage large landowners, like ranchers and farmers, to restore or retain large tracts of native grasslands by replacing perverse subsidies with payments for the ecosystem services that their unaltered lands provide ($232–$539 per hectare).

While there is no direct research attempting to assign value to the effects of prairie dogs on ecosystem services in natural grasslands, prairie dogs are strong and influential species in grassland ecosystems, meaning that they have the potential to drive or heavily influence deliverance of grassland services. Using information on the ecological role of prairie dogs in grasslands we hypothesize that prairie dogs have the potential to affect a wide variety of ecosystem services on grasslands (Table 7.4). Studies show increased nutrient cycling on prairie dog towns, which may affect carbon, nitrous oxide, and methane storage rates. Prairie dogs support the maintenance of grassland biodiversity by directly promoting the existence of nine grassland species and indirectly facilitating more than 100 species. In addition, burrowing activities that aerate soils on prairie dog towns may increase water infiltration rates. Grazing and burrowing activities that alter vegetation in relation to areas with no prairie dogs (namely increasing forb production, increasing native plant composition, and decreasing woody species) increases nutrient cycling and primary productivity rates, which could result in an increase in sequestration capacity and soil formation, as well as changes in pollinator composition and ground reflectance although this has not been documented. Changing ground reflectance and vegetation composition could in turn affect local temperatures and precipitation patterns. Thus, prairie dogs likely contribute quite strongly to grassland services and therefore should be considered to have high economic value in this regard.

In the first article to address the connection between species and ecosystem services in a direct way, Ehrlich and Mooney (1983) state: "The loss of services to humanity following extinctions ranges from trivial to catastrophic, depending on the number of elements (populations, species, guilds) deleted and the degree of control each exerted in the system." Prairie dogs, as keystone species, exert a high degree of control over grassland systems. In addition, their decline is linked to the decline of other elements on the population, species, and guild levels. There are no explicit economic values yet assigned to the proportion of grassland services that prairie dogs control. Yet, it remains clear that there is a risk to human society in lost grassland services if prairie dogs are removed and, because of their keystone status, the risk is potentially enormous.

In considering the issue of prairie dog management it is essential to consider the economics involved in prairie dog conservation. Externalities, or costs to the public from the devastation of prairie dogs and their habitats, are not properly factored into ventures that contribute to the decline of prairie dogs (ranching, agriculture, development, shooting). Because these externalities are not accounted for, decline of the prairie dog ecosystem persists at an amazingly fast rate because production costs to ranchers, farmers, developers, and recreational shooters are artificially low.

We conclude that, of all the various ways discussed in this chapter to obtain value of prairie dogs in an economic sense, the most promising is to evaluate them in terms of their contribution to grassland services. This technique is appropriate to prairie dogs because they are such large players in determining native grassland function. In addition, this technique connects prairie dog value with land values and so can easily be used to estimate values per hectare. This gives us a better way of evaluating the costs and benefits of our land conversion choices. It also provides a clear avenue to promote prairie dog conservation, for example by paying landowners a certain amount per hectare to give them the opportunity to keep or restore grasslands and the prairie dogs inhabiting them. Alternatively, these values per hectare could be used to charge landowners that choose to convert grasslands to other uses of marginal production value.

Quantifying the costs and benefits of prairie dog conservation enters prairie dogs into the dialog of our modern economy and lends more decision power to current conservation efforts. As the human footprint increases, we have become more acutely aware of how valuable natural entities are to us as a society and how underrepresented they are in our modern economy. We have seen that our current market does not adequately address or value issues of sustainability, natural health, and natural beauty. Environmental economics provides us with tools to reform our current economic system and to assure that externalities are truly accounted for. We must be more vigilant in forming our markets and judging our success as a society so that we integrate our ethical, cultural, and biological heritage values with our economic well-being.

We caution that assigning value, and attempting to include some living entities and ecological processes in an economic dialog is partly a quick fix that ignores a larger underlying problem about the inherent structure of our society as a whole. Currently our society bases success on a bottom economic line that includes economic growth (i.e. Gross Domestic Product), yet ignores many of the consequences of that economic growth. There are alternative ways of structuring societies and measuring the successes of societies that have the potential to foster sustainability. For example, there are alternative indicators that measure sustainability (Lawn 2003; Costanza 1996; Hanley et al. 1999; Stymne and Jackson 2000; Woodward 2000; Prugh and Assadourian 2003) and Emergy analysis (Odum 1996; Brown and Herendeen 1996; Brown and Ulgiati 1997). The topic of alternatives to purely economic drivers of political and social choices is large and controversial and cannot be fully dealt with in this book. However, it is an extremely important avenue of thought for us to consider in our quest for living in harmony with prairie dogs, and ultimately in living more sustainably on earth. Because, ultimately, as Morowitz (1991) aptly states, the real answer to the question, how much is a species worth? hinges on another question, what kind of world do we want to live in?

Interlude:
Prairie Dogs as Pets

Perhaps the earliest record of a pet prairie dog dates back to 1805 when Lewis and Clark sent President Thomas Jefferson a live black-tailed prairie dog, one of many specimens from their expedition. Whether Jefferson kept this prairie dog is unclear. What is certain, however, is that prairie dogs are currently a popular exotic pet. This trend is not limited to the United States. People all over the world, particularly in Japan, own prairie dogs as pets.

Until recently, there was relatively little regulation regarding the sale and export of wild prairie dogs for the pet industry. In May 2003 everything changed. The Centers for Disease Control (CDC) and the Food and Drug Administration (FDA) implemented a temporary ban on the sale and transport of wild prairie dogs in response to a monkeypox outbreak in May 2003 (USHHS, CDC, FDA 2003a).This outbreak was traced to a Texas animal distributor who imported a shipment of approximately 800 small mammals from Ghana on April 9, 2003. From that shipment, 584 of the original 762 African rodents were traced to distributors in six states. Investigations and CDC testing concluded that several rodent species were infected with the monkeypox virus. Apparently, prairie dogs that were housed near infected rodents in pet stores became ill with monkeypox. After an initial emergency ban in July 2003, an interim ruling was issued in November 2003 continuing the ban on sales, transportation, or any public or commercial distribution of prairie dogs. Specifically, the interim final rule was expanded to prohibit the "capture, offers to capture, transport, offers to transport, sale, barter, or exchange, offers to sell, barter, or exchange distribution, offers to distribute, or release of a listed animal into the environment regardless of whether the activity is interstate or intrastate" (USHHS, CDC, FDA 2003b). Currently, the ban is still enforced (USHHS, FDA 2005).

What does it mean to own a prairie dog? Is it like any other wild animal? Are they difficult to manage and care for? Or are they sweet, lovable, cuddly animals that make wonderful pets? The answer seems to depend on whom you ask. For a few, owning a prairie dog creates a dedication to learning about these animals, a commitment to preserving them in the wild, and a desire to educate the public about the care they require in captivity. For some, the joy and love they experience and feel for their prairie dogs is immeasurable. The bond shared between owner and prairie dog sometimes is wholly unexpected and beyond what individuals believed was possible. There is a general sense of surprise that prairie dogs are not as similar to hamsters or other rodents as may have been expected. In the wild, prairie dogs are social and gregarious, so perhaps it is not so unlikely that they exhibit a similar inclination toward sociability in captivity.

Oftentimes, a person's pet becomes like another member of the family. We see our pets as companions and we may rely on their presence when we are faced with difficulties. Whether it is a divorce, an illness, or other traumatic period in our life, we turn to our pets for comfort. Many of us view our relationships with them as reaching the depth, intensity, and level of emotional connection of any human relationship. When a pet is lost, the grief that is felt reflects the profound sense of love and connection we feel with them, and prairie dogs are no exception.

By and large, however, exotic pets are more difficult to take care of, handle, and bond with than a dog or a cat. One reason is that many wild animals require a special diet, enclosure, and environment in which to thrive. While there are people who undoubtedly have had wonderful experiences with prairie dogs as pets, there are others who have had little success in interacting with their new "family" members because they lack the basic knowledge of how they lived in the wild and how to care for them as pets.

As with other species, domestic and wild, there are shelters for abused, neglected, and abandoned prairie dogs. Pet prairie dogs have been burned with cigarette butts, confined and ignored in a small cage, hit, deposited in animal shelters and veterinarian offices where they are euthanized. Some are simply released outdoors. This is what happened to a female black-tailed prairie dog in New York. A lone prairie dog was discovered living in Robert Moses State Park. By August 2003, she had been living there for almost one year (Smith 2003). She had dug approximately eight burrows and seemed to be doing well and she provided entertainment to the children on the school bus that passed by every day. For this prairie dog there was a happy ending. She was trapped and sent to live with a dedicated and caring wildlife rehabilitator. For a fortunate few, this is the outcome, while for many the ending is grim.

There are medical challenges that seem to plague the captive prairie dog. Aside from monkey pox, which is not a typical disease of prairie dogs, owners must be aware of several diseases and conditions. One is tularemia, an infectious disease caused by the hardy bacterium *Francisella tularensis.* In July 2002, there was an outbreak of tularemia in Texas (JAVMA 2002). More worrisome is the prevalence of odontoma in captive prairie dogs. Odontoma is a type of cancer affecting the sinus area of the upper jaw. A tumor obstructs the flow of air through the sinus and is typically not curable. While there is disagreement on the exact cause of this disease in captive prairie dogs, trauma from chewing on cage bars is the leading suspect. Odontoma in prairie dogs is the most common type of illness seen in veterinary clinics and is the leading cause of death for pet prairie dogs. Obesity and respiratory illness also abound in pet prairie dogs, mostly because of lack of proper care.

A common theme among dedicated prairie dog owners is the notion that one is aiding in the survival of the species by having a prairie dog. The idea is that by owning a prairie dog it is saved from death in the wild, by either natural predators or human caused mortality. As attractive as this logic sounds, it is faulty for a number of reasons. First and foremost, most male pet prairie dogs are neutered, rendering them incapable of ever participating in the perpetuation of their species. Second, almost all prairie dogs in the pet trade are wild caught, effectively diminishing their number even further. It is difficult to determine exactly how many prairie dogs were captured annually prior to the ban, but the estimated number is large. One report states that in a single day 150 baby prairie dogs can be captured. People involved in the capture of wild prairie dogs are reported to receive $25–$30 per prairie dog and can earn $25,000–$30,000 a year, which would require the capture of some 1,000 prairie dogs. It is estimated that prior to the ban, 12,000 prairie dogs were exported out of the United States per year (Kuffner 2002). A single pet shop in the United States could sell as many as 3,000–5,000 a year (USHHS, FDA 2003). At one time, adults were captured and sold to pharmaceutical companies for gall bladder research (Moore 2002).

There are many individuals who are distressed by the current interim ban imposed by the FDA, and the reasons vary. For pet traders and store owners, it is an issue of profit. For those who love and care for their pet prairie dogs, quite simply, they cannot imagine a life without them. Yet, there is no automatic right to own an animal as a pet simply because it lives in our country. Most wild animals do best when left in the wild, and in most cases, prairie dogs are no exception.

8

Prairie Dog Conservation

Conservation biology was born in response to the widespread alarm of scientists, conservationists, foresters, and other resource managers that the world's biotic richness is declining at a sharply accelerating rate. The word *biodiversity* was coined to give a concrete name and coherent vision around which to organize conservation efforts. Conservation biology crosses a number of different disciplines in an effort to deal effectively with the large scope of the conservation dilemma. Conservation biology looks at ecological and evolutionary aspects of a problem and tries to offer solutions and insight that land and resource managers can use. Recently, conservation biology has increasingly incorporated social sciences into applied problem-solving in recognition that human social aspects are the primary drivers of the conservation problem. In this chapter, we build on information from previous chapters on biology, ecology, population biology, and economics to generate a clearer picture of the stresses and risks involved in prairie dog decline. We also examine the current status of all five species of prairie dogs and discuss possible tools that can be applied toward their conservation.

The Conservation Problem

A century ago, a person traveling through the American West would have reported seeing billions of prairie dogs in towns extending for hundreds of square miles. By 1960 black-tailed prairie dog colonies had been reduced to 600,000 hectares (about 1.5 million acres), less than 5% of their former range (Marsh 1984; Oldenmyer et al. 1993). Today the five species of prairie dog are continuing to decline rapidly. There are four major threats to the survival of prairie dogs: poisoning programs, shooting, habitat loss, and disease (Miller and Reading 2006). Three of these—poisoning, shooting, and habitat loss—are directly linked to humans. Even the fourth threat, disease, is linked to humans indirectly because people introduced plague into North America.

Poisoning of Prairie Dogs

For more than 100 years, federal, state, and local governments have subsidized and sponsored prairie dog poisoning programs to provide perceived relief to the livestock and agricultural industry. Prairie dogs were, and in many places still are, categorized as pests or varmints. People were—and sometimes still are—encouraged to poison prairie dogs. Following C. Hart Merriam's categorization of prairie dogs as a serious evil and his statement that 256 prairie dogs can eat as much grass as a single cow (a statement that seems to have lacked any evidence to support it) (Merriam 1902), the federal government began to get involved in poisoning. In 1917 the U.S. Department of Agriculture began a campaign to control prairie dogs, and in 1920 alone the Biological Survey employed 132,000 men to poison prairie dogs and ground squirrels on 13 million hectares (32 million acres) (Rosmarino 2003, personal communication). In 1931 the Animal Damage Control Act was passed, authorizing the eradication and control of animals such as the prairie dog. Poisoning continued from the 1930s through the 1950s. Meanwhile, starting in the early 1900s, states such as Texas, Kansas, and Colorado began their own poisoning programs. By 1911, the state of Kansas had eliminated prairie dogs from some 809,000 hectares (2 million acres), and between 1903 and 1911 the state of Colorado eradicated an estimated 91% of its prairie dogs (Forrest and Luchsinger 2004).

By 1960 black-tailed prairie dogs had been reduced to an estimated 600,000 hectares (1,482,000 acres), representing less than 5% of their historical range in the 1800s (Forrest and Luchsinger 2004). From 1975 to 1979, the Colorado Department of Agriculture poisoned prairie dogs on 253,000 hectares (625,184 acres), using Compound 1080, a poison that was later banned because of its extensive damage to other animals in an ecosystem. Poisoning campaigns continued through the 1980s and 1990s. In 1986 poisoning responsibility was transferred to the U.S. Department of Agriculture's Animal Plant Health Inspection Service (APHIS) program, to a group called Animal Damage Control, which has been recently renamed Wildlife Services. During the 1990s, it is estimated that Animal Damage Control poisoned an average of 40,000 hectares (100,000 acres) annually (Rosmarino 2003, personal communication). Animal Damage Control poisoned 13,218 hectares (33,045 acres) of prairie dogs in 1990 and 11,000 hectares (27,500 acres) in 1991 (U.S. APHIS ADC 1990; U.S. APHIS ADC 1991). Historically, the U.S. Forest Service, the U.S. Bureau of Land Management, the U.S. National Park Service, the U.S. Fish and Wildlife Service, the U.S. Bureau of Indian Affairs, the U.S. Animal Damage Control, and various state agencies have poisoned prairie dogs across their range. National forests across the Northern Great Plains usually delegate less than 5% of their available grassland to prairie dogs. When prairie dogs exceed this amount, they are targeted for elimination (Roemer and Forrest 1996). Today, Wildlife Services contracts with local governments and other parties to poison prairie dogs for a daily fee that includes the cost of the poison.

Several different poisons (or "toxicants") have been used. Originally poisons such as strychnine, thallium sulfide, compound 1080 (sodium fluoroacetate), and cyanide were used. These are currently banned because of their toxicity to other animals because of both secondary and non-target poisoning. Secondary poisoning results from scavengers eating the poisoned prairie dogs, and non-target poisoning results from animals other than prairie dogs eating the poisoned bait. Zinc phosphide is a commonly used poison today, and aluminum phosphide is used in urban regions (Forrest and Luchsinger 2004). Gas cartridges are also commonly used. For the gas cartridges, all the burrow openings are plugged first, and then gas cartridges are thrown down one burrow opening. The gas fills the burrow system and suffocates the prairie dogs (Hygnstrom and VerCauteren 2000). Unfortunately, these poisons affect not only prairie dogs, but can also cause poisoning or indirect mortality to other wildlife, such as burrowing owls, rabbits, snakes, lizards, and invertebrates.

Funding for prairie dog poisoning campaigns on public land comes from several sources (see chapter 7), including general appropriated funds, a small portion contributed by permittees (ranchers who hold grazing allotments and farmers leasing public land), and funds earmarked for conservation (Roemer and Forrest 1996). Typically, wildlife within National Parks cannot be targeted. However, exceptions regularly have been made with respect to prairie dogs, given their categorization as a pest. From 1982 to 1992, poisoning was regularly conducted in four national parks or monuments: Badlands National Park, Wind Cave National Park, Theodore Roosevelt National Park, and Devils Tower National Monument (U.S. NPS 1992). Funding for the poisoning of prairie dogs in National Parks was provided by the general operations budget (Cables 1993). Collectively from 1978 to 1992, roughly $10 million was spent on killing prairie dogs, not including additional taxpayer monies spent at the state and local level (Roemer and Forrest 1996). The agricultural and livestock industries have benefited directly from the federal, state, and local prairie dog control programs implemented to assist them. The livestock industry alone has enormous influence on the practices and regulations impacting their trade—in the past seven election cycles they have spent over $20 million lobbying Congress (Center for Responsive Politics 2003).

What of the notion that prairie dogs rob cattle of available food? One study suggests that prairie dogs reduce grazing biomass for cattle by as little as 4% (Uresk and Paulson 1988). Other studies have provided evidence that the forage quality, productivity, and nutritional value of grazing land are higher due to prairie dog activities (O'Meilia et al. 1982; Coppock et al. 1983a). While there may be a loss of forage that is available to cattle, this loss may be partially offset by an increase in forage quality (Detling 2004). Perhaps the perception that prairie dogs eat up all the grass in a given area comes from places like cattle ponds and other areas where cattle congregate. By assembling in a particular place, the cattle eat all the available vegetation and at this point prairie dogs often move in, for several reasons. One is that a lack of vegetation offers them good visibility for detecting predators. Another is that they have access to bare ground and to the seed bank in the soil. By eating high-energy seeds, the prairie dogs can potentially increase the amount of energy that is channeled to their reproduction (see chapter 3). However, when a rancher looks at the land, all he or she sees is a landscape denuded of vegetation and a lot of prairie dogs. So, the rancher often concludes that the prairie dogs stripped the ground bare, rather than concluding that the prairie dogs are there because of the cows' overgrazing. Prairie dogs are known to occupy highly disturbed sites such as around watering holes and homesteads (Knowles 1986). Their density tends to be lower in undisturbed grasslands (Travis and Slobodchikoff 1993). Through their selective feeding and clipping of plants, prairie dogs can prevent brush and introduced plants from becoming established, and can help maintain habitats as grasslands (Slobodchikoff et al. 1988; Weltzin 1997a, 1997b). Tens of millions of bison, whose diet is similar to that of cattle, have coexisted with black-tailed prairie dogs for thousands of years (Schwartz and Ellis 1981; Miller et al. 1994), yet the notion persists that prairie dogs cannot coexist with cattle.

Shooting Prairie Dogs

The second major threat to prairie dogs is shooting. As with poisoning, the deleterious effects of shooting are not limited to prairie dogs. Prairie dogs are near the bottom of the food chain, with many predators feeding on them. From badgers to mountain lions to eagles, prairie dogs help support a network of predators. Low densities of prairie dogs can cause a corresponding decline in predator populations (Knowles and Knowles 1994). The large-scale prairie dog population reduction is the leading cause for the extinction of the black-footed ferret in the wild. Also, secondary lead poisoning may be a problem for species that scavenge on dead prairie dogs (Pauli and Buskirk 2007). Ingestion of lead can be a problem for birds, and may affect burrowing owls, ferruginous hawks, golden eagles, and the recently re-introduced endangered California condor and black-footed ferret.

The reintroduction of the black-footed ferret is important to the discussion of prairie dogs because the success of reintroduction efforts is linked to the presence and persistence of large, healthy, stable populations of prairie dogs. In the grasslands of North America, the ferret's niche is one of a specialist predator. Aside from needing the extensive burrow system of prairie dogs as habitat, the black-footed ferret feeds almost exclusively on prairie dogs. The range of black-footed ferrets once corresponded with the historic range of prairie dogs, extending from Canada to Mexico (Cully 1993). As the range of the prairie dog was narrowed, so too was the ferret's. The black-footed ferret was afforded protection by the U.S. Fish and Wildlife Service in 1964 (Linder et al. 1972). Later, in 1973, ferrets were protected under the Endangered Species Act. At the same time, biologists recognized that ferrets depended heavily on prairie dogs for survival (Schroeder 1988). Section 7(a)(2) of the Endangered Species Act requires that federal agencies insure that any activity conducted is unlikely to negatively impact the survival of any threatened or endangered species, yet federal poisoning programs and recreational shooting of prairie dogs have persisted until the present. These factors are curtailing efforts to achieve a stable population of ferrets in the wild. In order for the reintroduction of ferrets to be successful, prairie dogs must be abundant and protected as a food resource. It is ironic that one branch of the federal government, the U.S. Fish and Wildlife Service, is spending millions of dollars to save the black-footed ferret, while another branch of the government, the USDA Wildlife Services, is spending millions of dollars to poison the principal food of the ferret.

The media and sport enthusiasts refer to the shooting of prairie dogs as a recreational sport. Many shooters believe they are conducting a public service by eliminating as many prairie dogs as possible. The dead prairie dogs are not consumed or used for any purpose—some shooters use exploding bullets to literally blow up the prairie dogs. Federal agencies, like the Bureau of Land Management, actively encourage the shooting of prairie dogs (Vosburgh and Irby 1998). During shooting, a larger number of females than males are killed, which causes a further decrease in the population, particularly if there are unweaned pups present. For example, one experimental study showed that after shooting in black-tailed prairie dog colonies, pregnancy rates declined by 50%, and the number of juveniles born per adult female declined by 76% (Pauli 2005). Contrary to popular belief, prairie dogs do not breed continuously (Hoogland 2001). All five species reproduce only once a year, with a mean litter size of 3.08–3.88 pups among black-tailed, Gunnison's, and Utah prairie dogs. Fewer than 60% of the pups survive through a single year. Shooting, particularly at the time that the females are lactating or the pups are just born, can severely impact the prairie dog population.

The scale at which shooting prairie dogs occurs is massive, especially considering such slow reproductive rates and low survivorship of offspring (Table 8.1). In 1998, in Conata Basin, South Dakota, 162,000 prairie dogs out of an estimated population of 216,000 animals were shot on 4,800 hectares (12,000 acres) (National Wildlife Federation 1998). In the years between 1993 and 2001 on the Lower Brule Sioux Reservation in South Dakota, an average of 14,200 prairie dogs were killed by an average of 121 shooters per year, with each shooter killing an average of 38 prairie dogs per day (Reeve and Vosburgh 2006). The hesitation of many states to prohibit shooting of prairie dogs may be linked to the proceeds funneled to state Game and Fish agencies through the Pittman-Robertson Wildlife Restoration Act, which allocates money to state wildlife agencies from an excise tax on guns, bullets, and other hunting equipment (Sharps 1988), as well as vocal opposition by some shooters and related businesses.

While shooting clearly reduces population densities, it also has additional side effects on prairie dog social behavior by reducing aboveground activity, thereby reducing the amount of time available for feeding (Vosburgh and Irby 1998). This could potentially increase stress levels, cause a decline in the overall health of the colony population, and make individuals more susceptible to disease. Additionally, colonies that are under shooting pressure have significantly reduced dispersal and expansion rates (Miller et al. 1993; Reading et al. 1989), thereby reducing migration, immigration, gene flow, and altering the population dynamics of prairie dog colonies.

Prairie Dogs and Plague

Another threat to the survival of the prairie dog comes from bubonic plague (see chapter 2). Plague can have a serious impact on prairie dog populations. With mortality rates of 99%, plague can cause the extirpation of entire colonies (Cully et al. 2006). A study of the effects of plague on Gunnison's prairie dogs in northern Arizona showed that out of 270 colonies that were known to exist in surveys conducted in 1987, between 1990 and 1994, and 1998, most (70%)

Table 8.1. Shooting estimates of prairie dogs from state agencies.

	1998	1999	2000	2001	2002
Arizona (Gpd)	—	—	91,864	75,791	21,134
Colorado (btpd/ Gpd/wtpd)	418,412	387,102	229,505	452,772	303,878
Kansas (btpd)	—	—	—	161,000	—
Nebraska (btdp)	301,000	365,000	203,600	—	—
Oklahoma (btpd)	—	Not reported	—	—	—
South Dakota (btpd) non-tribal only	—	1,425,437	1,246,862	1,520,000	—
Utah (UT pd only)	717	1,233	1,386	1,626	1,746

Source: E. Robertson, Center for Native Ecosystems.
Notes: Gpd = Gunnison's prairie dog, btpd = black-tailed prairie dog, wtpd = white-tailed prairie dog, UT pd = Utah prairie dog.

were inactive by 2000–2001 due to plague epizootics (the animal equivalent of epidemics in humans) (Girard et al. 2004; Wagner et al. 2006). Plague was responsible for an almost 100% die-off of prairie dogs on 250,000 hectares (625,000 acres) in South Park, Colorado, over a two-year period around 1950 (Ecke and Johnson 1952). In the Moreno Valley of northern New Mexico, prairie dogs went from being abundant in 1984 to small, isolated colonies in 1997, as a result of plague. In white-tailed prairie dog populations, plague appears to have somewhat less of an impact. Although white-tails are just as susceptible to plague, epizootics appear to happen less frequently, perhaps because the white-tails tend to occur at lower densities (Gasper and Watson 2001). The devastating effects of plagues are so visible on prairie dogs colonies that many people erroneously believe prairie dogs are responsible for the transmission of plague. In fact, because they die so rapidly after the onset of a plague epidemic, prairie dogs have rarely been the culprit of transmission of plague to humans (Barnes 1982).

Habitat Destruction

Along with other species that live in grasslands, prairie dogs are feeling the pressure of habitat destruction and encroachment by humans. Land development is subdividing and fragmenting already small populations of prairie dogs, compounding the effects of poisoning, shooting, and plague. While historically prairie dogs conflicted with humans in the areas of ranching and farming, a new conflict has emerged with the burgeoning population of people settling in the West. Many people move to places such as Colorado, Arizona, New Mexico, and Utah for the peace, serenity, and natural beauty of the "untamed" West. As more people move in to these areas, there are problems of urban sprawl from housing developments, roads, public buildings, and shopping malls. Prairie dogs live in grassy flat areas, and these are ideal for development purposes. Such development is reducing prairie dog habitat, fragmenting colonies, and undermining the survival of prairie dogs.

Since development seems inevitable as the demographic distribution of people shifts to more western states, the problem has been what to do with the prairie dogs. In recent years, this problem was addressed by poisoning the prairie dogs, or by simply burying the animals with bulldozers. A new trend has been to relocate prairie dogs to a safer habitat. Colorado was one of the first states to see a rise in public concern over the fate of prairie dogs on land slated for development. Organizations of volunteers arose out of this concern, promoting the welfare of prairie dogs and, if necessary, relocating them.

Relocations, or translocations, are often difficult and expensive (Truett et al. 2001; Long et al. 2006). Animals either have to be moved to a site that was previously occupied by prairie dogs and abandoned, or to a site where artificial burrows and nest boxes are dug for the prairie dogs. If the animals are moved to the site of an abandoned colony, there is the possibility of residual plague that remains among other less-susceptible species, or among fleas, that could kill the translocated animals (most abandoned colonies have no prairie dogs because plague killed those that were there). If the animals are moved to a site that is on public land leased for cattle grazing from a federal or state agency, the permission of the permitee or leasee must be obtained before any animals can be translocated. Criteria for the release of Utah prairie dogs require a 1.6-kilometer (1-mile) distance or some kind of structural barrier between the release site and private land, while in New Mexico even more stringent criteria must be used to prevent black-tailed prairie dogs from moving onto neighboring land.

The politics of moving animals can be difficult. In 1999 Colorado passed a law (SB 99–111) that prohibits moving prairie dogs across county lines without the express permission of the county commissioners. Once the prairie dogs are moved to a new site, some counties and states levy fines on the property owners if even a single prairie dog strays into a neighboring property.

In translocating prairie dogs, the animals have to be removed from their former colony and then monitored at the new site. This requires considerable time and effort. Two common ways of removing animals are by trapping and by flushing. In the trapping method, the social groups of the animals have to be identified first, then traps have to be pre-baited so that the animals will be accustomed to them, and then the animals have to be trapped and moved according to their social groups. In the flushing method, a combination of water and soap is used to flush prairie dogs out of their burrows. As with trapping, flushing can also be used to move entire social groups. Once they are dried, the prairie dogs are moved to new burrows at the translocation site. This method requires a large quantity of water (Truett et al. 2001), and is more appropriate for urban areas where water is readily available.

Monitoring animals at the new site is usually done with visual counts of prairie dogs (Severson and Plumb 1998). Success is improved if the prairie dogs are temporarily held on a site in small, predator-proof enclosures and provided with food.

Despite these difficulties, translocation offers a humane alternative to killing prairie dogs. At present, much of the work is being done by volunteers, with voluntary cooperation by developers and state and local agencies. However, some governments are now mandating that developers consider translocation as a first alternative to solving their prairie dog "problem." For example, the City of Santa Fe, New Mexico, passed an ordinance (Bill No. 2001–37) in 2001 requiring that Gunnison's prairie dogs be humanely relocated by a certified relocator to appropriate and safe areas prior to the onset of development on the site of a colony. This, however, is only a temporary fix because there are only so many places prairie dogs can be moved. A better fix is growth planning that includes open spaces where prairie dogs can continue to live, and acquisition of land where prairie dogs can be transplanted.

Counting Prairie Dogs

A significant problem in the conservation of prairie dogs is determining how many there are in any given area. Someone can go to a colony and count how many animals are visible, but this is not a reliable number because many prairie dogs spend part of their day belowground, and the number aboveground at any point can be highly variable. Several methods have been employed, and none of them is completely reliable (Slobodchikoff et al. 1988; Powell et al. 1994; Severson and Plumb 1998; Johnson and Collinge 2004; Biggins et al. 2006) Visual counts are one method. Another method is trapping and marking all of the animals in a colony. This is also not entirely reliable, because not all of the animals can be trapped (Biggins et al. 2006). Juvenile animals are often shy about going into traps, and some adult animals refuse to go into a trap regardless of the incentives that are offered to them. Even were this method completely effective, as it sometimes is with smaller colonies, it is impractical to capture and mark every single prairie dog in hundreds of colonies. A third method is to use transects from low-flying aircraft, where an observer marks the start and the end of a colony from the burrow openings, and the colony area is calculated from a series of transects. Yet another method is to use either aerial photographs or satellite imagery to calculate areas containing prairie dog burrow openings. These methods cannot distinguish between active prairie dog burrows and burrows that have been abandoned because the prairie dogs have either died of plague or have been poisoned. Although the newer satellite technology, such as Ikonos imagery with 1-meter resolution, can give very precise locations of burrow openings, some measure of ground-truthing is necessary to establish whether or not prairie dogs are actively using the burrows (Biggins et al. 2006).

Counting burrows poses another problem, how does the number of burrow openings translate into the number of prairie dogs? In a study of a colony of Gunnison's prairie dogs where all of the individual animals residing in the colony were marked, Slobodchikoff et al. (1988) found that burrow openings and prairie dogs occurred in a 2:1 ratio, that is, there were 2 burrow openings for every prairie dog. Other studies have found that there was an average of 3.9 burrow openings per black-tailed prairie dog (Biggins et al. 2006), and an average of 11.5 burrow openings per white-tailed prairie dog (Biggins et al. 1993). However, some studies of black-tailed prairie dog colonies did not find any significant correlation between the number of burrow openings and the number of prairie dogs in a colony (Powell et al. 1994; Severson and Plumb 1998). Determining an accurate way to relate burrow openings to prairie dog numbers remains a priority for conservation efforts.

Prairie Dogs, the Endangered Species Act, and CITES

The Endangered Species Act offers a potential avenue for reversing prairie dog decline. In 1973 the United States Congress declared that "various species of fish, wildlife, and plants in the United States have been rendered extinct as a consequence of economic growth and development untempered by adequate concern and conservation" and that "these species of fish, wildlife, and plants are of esthetic, ecological, educational, historical, recreational, and scientific value to the Nation and its people." This declaration became part of the Endangered Species Act (16 U.S.C. 1531–1543), whose purpose is to "provide a means whereby the ecosystem upon which endangered species and threatened species depend may be conserved, to provide a program for the conservation of such endangered species and threatened species, and to take such steps as may be appropriate to achieve the purposes of the treaties and conventions set forth . . ." The Endangered Species Act replaced the Endangered Species Preservation Act of 1966, which required a list of endangered species but provided no regulations that prevented the killing of such species. Also replaced was the Endangered Species Conservation Act of 1969, which mandated establishing a list of endangered species that were foreign to the United States. The Endangered Species Act defines an endangered species as "any species which is in danger of extinction throughout all or a significant portion of its range . . ." and defines a threatened species as "any species which is likely to become an endangered species within the foreseeable future throughout all or a significant portion of its range." Similar to the Endangered Species Act is the Convention on International Trade in Endangered Species of Wild Fauna and Flora (CITES), a United Nations program, which attempts to protect endangered species worldwide.

The Endangered Species Act, CITES, and several opinion polls conducted through the years (including one in 2002) make it clear that we the public have been, and still are, concerned with the persistence of species, the maintenance of biodiversity, and the preservation of ecosystems on the planet. As a result of the Endangered Species Act and CITES, species that would have vanished forever were kept from the brink of extinction. The bald eagle continues to soar gracefully across the skies of America, the wolf is once again roaming Yellowstone National Park together with the American bison, and the California condor is presently on its way back into the wild.

Where do prairie dogs stand, in terms of the Endangered Species Act and CITES? On June 2, 1970, the Mexican prairie dog was designated as endangered in its entire range, Mexico, under the provisions of the Endangered Species Preservation Act (Federal Register 35 FR 8491–8498). On January 7, 1975, the Mexican prairie dogs was listed as endangered in CITES. In 2000 the Mexican prairie dog was placed on the International Union for Conservation of Nature and Natural Resources (IUCN) Red List as endangered. The IUCN Red List follows specific criteria to determine what category a species should fall into and these categories range from extinct to lower risk. A species that is Red Listed is endangered and facing a very high risk of extinction in the wild under the any of the following criteria (IUCN Red List Categories and Criteria Version 2.3):

- A population reduction of at least 50% over the last 10 years or three generations

- A projected population reduction of at least 50% over the next 10 years
- The range over which a species occurs is less than 5,000 km² or the species occupies an area less than 500 km²
- The population contains fewer than 2,500 mature individuals with subpopulations containing fewer than 250 mature individuals or extreme fragmentation between populations, none of which contain more than 250 mature individuals
- The population is estimated to be fewer than 250 mature individuals
- If extinction simulations reveal the probability of extinction is 20% within 20 years or five generations

At almost the same time as the Mexican prairie dog, the Utah prairie dog was designated on June 4, 1973, as endangered over its entire range, Utah, by the Bureau of Sport Fisheries and Wildlife (the precursor of the U.S. Fish and Wildlife Service) (Federal Register 38 FR 14678). Prior to that, it had been classified as endangered in 1968 then reclassified as non-endangered in 1972 (Collier and Spillett 1972). In 1979 the Utah Division of Wildlife Resources petitioned the U.S. Fish and Wildlife Service to remove the Utah prairie dog from the list of endangered species. As a result, the Utah prairie dog was reclassified as threatened on May 29, 1984 (49 FR 22330–22334), a designation that it retains at present.

As part of the reclassification from endangered to threatened in 1984, a special rule was implemented to allow the "take" of Utah prairie dogs (50 CFR 17.40(g)). The number of Utah prairie dogs that could be "taken" was limited to 5,000 animals annually (USFWS 1991). The Endangered Species Act defines "take" to mean: to harass, harm, pursue, hunt, shoot, wound, kill, trap, capture, or collect, or to attempt to engage in any such conduct. In 1988 there was an estimated total population of 5,984 Utah prairie dogs, and in 1989, with 1,267 prairie dogs killed, the population was estimated to be 7,377. In 1991 the "take" was increased to 6,000 individuals and is still in effect as of June 2008 (56 FR 27438–27443). The census count of adult Utah prairie dogs in 2007 was 5,991 individuals, which might represent only 50% of the total population size (Rosmarino, personal communication). It appears that under its protected status of "threatened," the annual allowed "take" of the Utah prairie dog may be greater than half its estimated population size.

The black-tailed prairie dog is not currently listed for federal protection. Two petitions were filed to list the black-tailed prairie dog on the Endangered Species List, both in 1998. The first was filed by the National Wildlife Federation and the second by the Biodiversity Legal Foundation, Predator Conservation Alliance, and John Sharps. As a result, on February 4, 2000, the black-tailed prairie dog was designated by the U.S. Fish and Wildlife Service as a Candidate Species, with a Warranted But Precluded listing (64 FR 53655–53656). Warranted But Precluded means that the U.S. Fish and Wildlife Service staff found sufficient reason to list the species but there is not sufficient money available to carry out the details of listing. As a Warranted But Precluded Species, the black-tail was given a priority rank of 8 (from a scale of 1–12, with 1 the highest priority for listing when funds become available) (66 FR 54807–54832; 67 FR 40657–40679). However, on August 12, 2004, the Interior Department announced that it was dropping the black-tailed prairie dog from all consideration for endangered or threatened status (69 FR 51217–51226), a decision that was hailed by the governors of several states and a variety of politicians who took credit for this action (Chet Brokaw, AP Writer, *Rapid City Journal*; Theo Stein, Denver Post Writer, *Denver Post*, August 13, 2004).

The decision to not list the black-tailed prairie dog was made primarily on a reevaluation of the area encompassed by prairie dog colonies (69 FR 51217–51226). A number of states produced new estimates of prairie dog populations, mainly through aerial surveys, and indicated that black-tailed prairie dog populations were larger than had previously been thought. In the year 2000, the U.S. Fish and Wildlife Service estimated that black-tailed prairie dogs were found on 311,000 hectares (777,500 acres) in North America (Manes 2006). In the year 2004, based on new estimates from state agencies, the U.S. Fish and Wildlife Service estimated that prairie dogs inhabited approximately 745,400 hectares (1,842,000 acres). They estimated that prairie dog densities average about 10 animals per acre, with a range between 2–18 animals per acre, and concluded that there are approximately 18,420,000 black-tailed prairie dogs in North America (69 FR 51217–51226). A big contributor to the size of this estimate was the state of Colorado. A 1961 estimate reported that Colorado had 39,000 hectares of prairie dogs (97,500 acres), and the 2000 estimate reported 38,000 hectares (95,000 acres) (Manes 2006). However, for the 2004 estimate, Colorado reported 256,000 hectares (631,000 acres) (69

FR 51217–51226), an increase of 674% in four years. Another big contributor to this estimate was the state of South Dakota, who reported 60,000 hectares (147,000 acres) of prairie dogs in 2000 and 165,000 hectares (407,000 acres) in 2004, an increase of 275% (69 FR 51217–51226). Other states reported fewer hectares of prairie dog populations (Manes 2006).

The accuracy of the Colorado estimate is subject to some question (Miller et al. 2005). The Colorado estimate was based on aerial transects that determined the points at which prairie dog digging or animals were observed along these transects (White et al. 2005). Miller et al. (2005) examined 18 of the 1,596 transects (3.9% of the total) used by White et al. (2005), walking along these transects as a way of "ground-truthing" the aerial observations. Miller et al. (2005) found that 75.7% of the total length of the transects that they walked did not have any prairie dog activity, with 50.3% having only inactive burrows left by prairie dogs that had been poisoned or succumbed to plague at some time in the past, and 25.4% of the length of these transects entirely lacking in any signs of prairie dog burrows.

Opinions differ as to whether the U.S. Fish and Wildlife Service made a biologically valid decision to remove the black-tailed prairie dog from the candidate species list. Manes (2006) has argued that the U.S. Fish and Wildlife Service made a good decision, because new information had shown that there are more black-tailed prairie dogs than people previously had thought existed. Additionally, Manes (2006) points out that the U.S. Fish and Wildlife Service decided that plague is not going to seriously impact black-tailed prairie dog populations, because the populations east of the 102nd meridian are mostly free of plague, there might be indications that some prairie dogs are developing a resistance to plague, and because colonies are smaller and more widely scattered today than they were in the past, there is less chance that plague will occur at any given colony. On the other hand, Rosmarino (2006) has argued that because prairie dogs suffer from threats of plague, poisoning, shooting, and loss of habitat, and because prairie dogs are keystone species in grassland ecosystems, policymakers should practice the "better safe than sorry" practice and list the black-tailed prairie dog as threatened before populations decline any further.

Regardless of merits of removing the black-tailed prairie dog from the candidate list, various states responded quickly by poisoning prairie dogs on large tracts of land. Hoogland (2006) points out that in 2004 after the removal of the black-tailed prairie dog from the candidate list, poisonings occurred at Conata Basin, Buffalo Gap National Grasslands in South Dakota, on 2,800 hectares (6,900 acres), and on 10,931 hectares (27,000 acres) of private land bordering federal land in South Dakota. The Conata Basin is one of the few sites where the endangered black-footed ferrets, whose principal food is prairie dogs, have been successfully reintroduced at great expense of time and taxpayer money. Hoogland (2006) also points out that in 2005, the South Dakota legislature passed two bills, HB-1252 that allows the poisoning of prairie dogs and provides funds for annual poisoning on private lands adjoining federal land, and SB-216, that classifies the black-tailed prairie dog as a pest. Also, in 2004 the South Dakota Department of Game, Fish, and Parks lifted its ban on recreational shooting of prairie dogs in Conata Basin, Buffalo Gap National Grasslands (Hoogland 2006).

Petitions to list other species of prairie dogs have not fared very well. A petition to list the white-tailed prairie dog was submitted to the U.S. Fish and Wildlife Service on July 15, 2002 by the Center for Native Ecosystems, Biodiversity Conservation Alliance, Southern Utah Wilderness Alliance, American Lands Alliance, WildEarth Guardians, the Ecology Center, Sinapu, and Terry Tempest Williams. On November 2, 2004, the U.S. Fish and Wildlife Service issued a 90-day ruling (a ruling that they are required to complete by law in 90 days, saying whether the petition has merit for further consideration), declaring that the white-tailed petition did not provide sufficient scientific or commercial information to warrant listing the species (FR 69: 64889–64901). The 90-day finding issued by U.S. Fish and Wildlife Service contains a table that lists population estimates of white-tailed prairie dog colonies that were considered large enough for possible introductions of black-footed ferrets. Out of nine sites, all but two of them had lower population estimates in 2001–2003 than in previous years (FR 69: 64889–64901):

Wyoming
Shirly Basin, 1991, 30,389 prairie dogs, 2001, 34,698 prairie dogs
Meeteetse, 1988, 25,494 prairie dogs, 2000, 1,066 prairie dogs

Colorado

Coyote Basin, 1997, 3,132 prairie dogs, 2003, 1,055 prairie dogs
Wolf Creek West, 2000, 19,719 prairie dogs, 2003, 9,214 prairie dogs
Wolf Creek East, 2001, 10,331 prairie dogs, 2003, 10,754 prairie dogs

Utah

Coyote Basin, 1997, 43,205 prairie dogs, 2003, 14,031 prairie dogs
Kennedy Wash, 1998, 10,697 prairie dogs, 2003, 3,313 prairie dogs
Shiner Basin, 1997, 15,065 prairie dogs, 2000, 13,707 prairie dogs
Snake John, 2001, 49,346 prairie dogs, 2003, 31,118 prairie dogs

The 90-day finding also has a table showing that in Montana, the total habitat occupied by 15 white-tailed colonies in 1979 was 289 hectares (690 acres), compared to 48 hectares (120 acres) occupied by 6 colonies in 1999–2003, a decline of 85%. Of these latter six colonies, two were sampled between 1975 and 1977. One colony went from an estimated 30–34 hectares (74–84 acres) to a size of 5.1 hectares (12 acres), and the other went from 100 hectares (250 acres) to a size of 16.4 hectares (40.5 acres) (FR 69: 64889–64901), an average decline of 84% in colony size in about 25 years. Part of the argument of the white-tailed petition was that white-tailed prairie dogs now occupy less than 322,000 hectares (805,000 acres), which represents a 92% reduction in their range in the past century. Another part of the argument was that white-tailed prairie dogs are being shot at a phenomenal rate. The Colorado Division of Wildlife estimates that between 1999 and 2001, some 65,000 to 122,000 white-tails were shot in Colorado (ESA White-tailed Prairie Dog Petition). However, in November of 2007, the U.S. Fish and Wildlife Service announced that they would reconsider the evidence for listing the white-tailed prairie dog.

Similarly, a petition to list the Gunnison's prairie dog was filed on February 23, 2004, by WildEarth Guardians and 73 other organizations and private individuals. The U.S. Fish and Wildlife Service issued a 90-day finding on January 30, 2006, saying that there was not sufficient evidence to pursue consideration of listing of the Gunnison's prairie dog (FR 71: 6241–6248). As with the white-tailed prairie dogs, the 90-day finding contains information on the estimated population sizes and estimates of colony numbers. Of six colonies for which population numbers are available, only one of the six (Aubrey Valley, Arizona) was increasing in 2005, and the other five were decreasing (three colonies) or extinct (two colonies).

Arizona
Aubrey Valley, 1990, 19,368 prairie dogs, 2005, 42,000 prairie dogs

Dilkon, 1994, 8,650 prairie dogs, 2001, 43 prairie dogs

Colorado
Currecanti National Recreational Area, Gunnison County, 1980, 148 prairie dogs, 1981, extinct

Saguache, Montrose County, 1980, 15,569 prairie dogs, 2002, 770 prairie dogs

South Park, 1945, 915,000 prairie dogs, 1948, 74,000 prairie dogs, 2002, 42 prairie dogs

New Mexico
Catron and Socorro Counties, 1916, 2,458,650 prairie dogs, 1984, less than 12,000, 2005, less than 6,000

Moreno Valley, 1984, 11,000 prairie dogs, 1987, extinct

The 90-day finding also lists sites where multiple colonies were known to have occurred, and of the six sites that are listed, all have decreased in colony numbers (FR 71: 6241–6248).

Arizona
Flagstaff, 2000, 75 colonies, 2001, 14 colonies

Petrified Forest National Park, 1994, 8 colonies, 1996, 3 colonies

Seligman, 1990, 47 colonies, 2001, 11 colonies

Colorado
Chubbs Park, 1958, 1 colony, 1959, extinct

New Mexico
Navajo Nation, 1966, 625 colonies, 1969, 233 colonies

Utah
Garfield County, 1980, 1 colony, 1981, extinct

In July of 2007, the U.S. Fish and Wildlife Service agreed to reconsider their 90-day finding, in response to legal action from a variety of environmental groups, and in 2008 found that Gunnison's prairie dogs warrant listing in the mountain portion of their range and were designated as a candidate species.

At the same time that U.S. Fish and Wildlife Service was considering the listing petitions, an 11-state management plan was beginning to be drafted in an effort to develop conservation guidelines for black-tailed prairie dogs (Van Pelt 1999, 2000; Luce et al. 2006). The problems of coming up with a conservation strategy, however, are formidable.

Part of the challenge of developing a unified management plan is that the states have different laws and different views about prairie dogs (Tables 8.2, 8.3, 8.4). For example, consider the variety of regulations concerning black-tailed prairie dogs. In Arizona, black-tailed prairie dogs were extirpated in the 1950s and currently do not occur in that state. In 1999 the Arizona Game and Fish Commission banned the shooting of black-tailed prairie dogs (of which there are none!). Arizona currently is exploring plans to re-introduce this species. In Colorado, the black-tailed prairie dog is classified as a small game species. The Colorado Division of Wildlife recently banned the shooting of prairie dogs on public lands, but shooting on private land and poisoning are still allowed. In 1999 the Colorado legislature passed a bill (SB 99–111) that prohibits anyone from moving prairie dogs across county lines without the permission of the county commissioners. In Kansas the black-tailed prairie dog is considered to be wildlife, requiring a hunting permit, but the season is open all year with no bag limits. In Montana black-tails were classified as vertebrate pests and could be killed by the Department of Agriculture. In 2000 the Montana state legislature passed a bill upgrading the black-tailed prairie dog to a nongame species, with provisions to ensure that no private landowners would be restricted from killing prairie dogs on their land.

In Nebraska the black-tailed prairie dog is not protected, and can be shot or poisoned without restrictions. Control activities are continuing to be implemented by landowners directly, by pest control agents, or with the assistance of U.S. Department of Agriculture Animal and Plant Health Inspection Service Wildlife Services (APHIS-WS). In New Mexico, there is no legal protection for prairie dogs. In North Dakota residents are not required to have a hunting license to shoot prairie dogs. There is no hunting season or bag limit. In Oklahoma the black-tailed prairie dog is classified as a Category II Mammal Species of Special Concern, which still allows prairie dogs to be poisoned or to be killed with rifles, shotguns, handguns, and bows and arrows.

In South Dakota the black-tailed prairie dog is designated as a game species and as a predator/varmint. Prairie dog shooters must have a license, but there is no limit on where, when, and how many prairie dogs can be shot. The South Dakota Department of Agriculture is legally empowered, at state expense, to poison prairie dogs that establish themselves on private property from neighboring public lands. In 2002 the South Dakota legislature mandated that no prairie dog management plan could be implemented without legislative approval. In Texas the Texas Parks and Wildlife Department listed black-tailed prairie dogs are listed as a nongame species, and a hunting license is required, but they can be shot all year and there is no bag limit. Poisoning of prairie dogs is also allowed. In Wyoming, the Wyoming Department of Agriculture classifies the black-tailed prairie dog as a pest, with no restrictions on the killing of prairie dogs. Although the Wyoming Game and Fish Department lists the black-tailed prairie dog as a Species of Special Concern, prairie dogs may be shot during the entire year without a permit.

Table 8.2.   Game status, agricultural status, and shooting regulations for black-tailed prairie dogs.

State	G&F status	AG status	Shooting regs.
Arizona	Non-game mammal/extirpated		Closed season
Colorado	Small game species	Destructive rodent pest	Closed on federal land/ permitted on state and private land
Kansas	Wildlife	Pests/mandatory control	No restrictions
Montana	Nongame wildlife species of special concern	Rodent and invertebrate pest	No restrictions except federal land closed March 1–May 31
Nebraska	Unprotected nongame wildlife	Pest	No restrictions
New Mexico	Nongame species	Rodent pest	No restrictions
Oklahoma	Category II species of special concern	Not pests, still poisoned	No restrictions
North Dakota	Nongame wildlife	Pest	No restrictions
South Dakota	Game wildlife species of management concern	Pest	No restrictions except on public lands from March 1 to June 14
Texas	Nongame wildlife	Pest	No restrictions
Wyoming	Nongame wildlife/species of special concern	Pest	No restrictions

Source: S. Nichols-Young 2003, Animal Defense League of Arizona.

Notes: "G&F status" refers to the designation by state wildlife departments, which have different names in different states.

"AG status" refers to the designation by state agriculture departments.

Conservation Tools: What Can Be Done to Keep Prairie Dogs from Going Extinct?

Several groups are attempting to provide a proactive approach to the conservation of prairie dogs. After the petition to list the black-tailed prairie dog was filed in 1998, representatives from 11 states began the process of putting together a multistate plan to conserve prairie dogs, forming the Interstate Black-tailed Prairie Dog Conservation Team (Luce et al. 2006). In 1999 representatives from

Table 8.3. Game status, agricultural status, and shooting regulations for white-tailed prairie dogs.

State	G&F status	AG status	Shooting regs.
Colorado	Small game species		No restrictions except closed on national wildlife refuges
Montana	Species of concern		No restrictions except closed on public lands
Utah	Nongame mammal, sensitive species		No restrictions on private lands; closed on public lands from April 1 to June 15
Wyoming	Nongame wildlife species	Pest	No restrictions

Sources: S. Nichols-Young 2003, Animal Defense League of Arizona; Seglund et al. 2006a.
Notes: "G&F status" refers to the designation by state wildlife departments, which have different names in different states.
"AG status" refers to the designation by state agriculture departments.

Table 8.4. Status and shooting restrictions of Gunnison's prairie dogs.

State	G&F status	Shooting regs.
Arizona	Nongame species	No restrictions except closed from April 1 to June 15 on public and private lands
Colorado	Small game species	No restrictions
New Mexico	Nongame species	No restrictions except closed on state trust lands and wildlife management areas
Utah	Nongame mammal, sensitive species	No restrictions on private lands; closed from April 1 to June 15 on public lands

Sources: S. Nichols-Young, 2003, Animal Defense League of Arizona; Seglund et al. 2006b.
Note: "G&F status" refers to the designation by state wildlife departments, which have different names in different states.

nine of those states (except for Colorado and North Dakota) signed a Memorandum of Understanding, pledging to cooperate in the efforts to conserve prairie dogs. Each state set up committees with people who had some interest in prairie dogs, such as farmers, cattlemen, shooters, landowners, and environmentalists. The plan developed by these committees calls for preserving the current prairie dog habitat, and trying to increase the habitat on which prairie dogs are found by 9% by the year 2011. Although this is an excellent step in the right direction, one problem is that 87% of the land on which prairie dogs are currently found is held in private ownership, and only 5% of the land occupied by prairie dogs is federally owned, with an additional 8% owned by Native American tribes (Luce et al. 2006). Private owners can disagree with state management agencies on the best policy to pursue with prairie dogs, and many private landowners want prairie dogs eliminated from their property, as seen in a survey in Wyoming where 54% of the private landowners wanted to eliminate prairie dogs from their land (Luce et al. 2006). Without the prospect of federal listing of prairie dogs, some state legislatures (e.g., South Dakota) are passing legislation legitimizing or making it easier to poison prairie dogs on private property. Also, without the prospect of federal listing, state agencies have less incentive to apply conservation strategies to the prairie dogs within their states.

Another group, the Southern Plains Land Trust, is taking the approach of buying land and trying to preserve the prairie dogs and other grassland species found on that land. The Southern Plains Land Trust purchased their first parcel of land, 512 hectares (1280 acres) in November 1998 in grassland near Comanche National Grassland, Colorado. Currently, approximately 1,296 hectares (3,200 acres) are enrolled in the trust's preserve network seeking to restore shortgrass prairie ecosystems in the Southern Plains.

A third group is trying to determine focal areas of conservation from maps, surveys, and records of prairie dog occurrence (Proctor et al. 2006). Focal areas are defined as places that are a minimum of 4,000 hectares (10,000 acres) and are large enough and diverse enough to contain functioning ecosystems, in other words the prairie dogs and all of the grassland species that depend on the prairie dogs. Proctor et al. (2006) have identified 84 focal areas that are large enough to support a functional grassland ecosystem with black-tailed prairie dogs, and most of these are on federal or state lands. Effort is placed on convincing federal and state agencies to restore and conserve black-tailed prairie dogs within these focal areas that represent a part of the historical range of the animals.

A fourth group is the Prairie Dog Coalition, a group that represents a consortium of environmental and conservation organizations that are trying to save prairie dogs. The Prairie Dog Coalition disseminates information about prairie dog-related activities, such as places where prairie dogs are in danger of being poisoned or eliminated, or legislation that affects the welfare of prairie dogs. The coalition sponsors conferences and events that bring together representatives from a number of organizations that have various interests in saving prairie dogs, ranging from organizations that are directly involved in transplanting prairie dogs from habitat that is scheduled to be developed, to organizations that are concerned with the preservation of grassland ecosystems.

A fifth group is a number of environmental organizations such as Wild Earth Guardians, Biodiversity Conservation Alliance, and Center for Native Ecosystems, that alert their members to threats to prairie dogs and other species that are potentially endangered, and also help craft petitions arguing for listing of declining species. When the federal government does not respond to petitions within the required deadlines (as for example, the U.S. Fish and Wildlife Service responding with a 90-day finding for the white-tailed prairie dog petition more than two years after the petition was filed), these organizations can take legal action, forcing a response from appropriate federal agencies.

Any attempt to successfully coexist with prairie dogs is not complete unless it addresses the intense social issues connected to prairie dog conservation. This requires bringing in all the people who have a stake in the life and death of prairie dogs—ranchers, farmers, agency land managers, scientists, animal rights groups, prairie dog shooters, and conservation organizations.

Several surveys have looked at the attitudes of different segments of the public toward prairie dogs (Lamb et al. 2006). One survey looked at the attitudes and values of 900 Montana residents toward prairie dogs (Reading et al. 1999). This group included 300 urban residents of Billings, Montana, 300 rural residents of Phillips County, 150 Montana ranchers, and 150 members of Montana non-governmental conservation organizations. Except for the members of the conservation organizations, all the other groups considered prairie dogs as pests and showed little concern for their welfare. Urban residents and conservation organization members indicated that they enjoyed watching prairie dogs, while rural residents and ranchers did not enjoy watching the animals. About 46% of the conservation organization members and 30% of urban residents said that they would like to see more than 5% of public grazing land set aside for prairie dogs, while 56% of the ranchers and 36% of rural residents indicated that no public grazing land should be set aside. Most people in the survey cited personal experience as their primary source of information about prairie dogs.

A similar survey was conducted in Fort Collins, Colorado (Zinn and Andelt 1999). This survey compared the attitudes of 87 residents living near prairie dog colonies with the attitudes of 559 randomly-sampled residents of the city. The results were that people living near prairie dog colonies tended to see prairie dogs as unattractive, as a useless pest, and saw poisoning as the best way to remove the animals. The general city population sample tended to see prairie dogs as more attractive, and favored relocation as a way of removing animals from an area. Both groups generally favored preserving some prairie dog colonies around the city. Few residents in either group knew that prairie dogs had only one litter per year (14% of those living near prairie dogs and 8% of those in the general sample), and a little more than one-third of either group knew that prairie dogs are less common now than they were in 1800 (36% of those living near prairie dogs and 43% of the general sample).

These results are not encouraging news for the conservation of prairie dogs. Conservation organizations can generate political support to change Federal, State, and local policies toward management and poisoning only if they have a broad base of support from the general public or have other sources of power and sway over politicians. These surveys suggest that this broad base of support does not yet exist. Also, because of their potential political power, ranching interests can influence government agencies more easily than can conservation organizations or members of the general public without a broad base of support.

Tackling the social side of prairie dog conservation will likely need to include education, and alleviation of political, social, and economic influences that perpetuate the mistreatment of prairie dogs. This is a daunting task but it is one that is mandatory if conservation efforts are to be successful. In fact, failing to deal with social aspects of conservation is emerging as the major reason for failure of conservation projects worldwide (Vogt et al. 2001).

The concerns of ranchers and farmers have to do with their ability to make a living, changes to their lifestyles, loss of control over public and private range lands, and similar issues. Practical solutions must be found that will allay those concerns. One solution is paying ranchers and farmers to have prairie dogs on their land, rather than paying for poisoning and control efforts (Miller et al. 1990). This is similar in concept to the conservation plan of Kenya in preserving its large animals where farmers are compensated for having large animals on their property. A proposal to do this was submitted to the U.S. Department of the Interior in 2001 by the Interstate Black-tailed Conservation Team. The proposal suggested paying landowners who were willing to conserve prairie dogs on their land a one-time payment of $250 per hectare ($100 per acre) and an average payment of $25 per hectare ($10 per acre) over the space of 10 years. The proposal was not accepted by the Interior Department (Luce et al. 2006).

Recreational shooters like to kill prairie dogs for the challenge presented by firing at something from a long distance (Reading et al. 1999). Perhaps some land can be set aside with pop-up targets at different distances, similar to a military firing range, so that shooters can experience the same challenge to their skill without actually killing any prairie dogs. Areas of grasslands without prairie dogs can be set aside for this purpose, and economic interests such as hotels and restaurants can be accorded space on the edges of such areas to induce shooters to come and stay at such places.

Agency land managers have to be persuaded to allow larger numbers of prairie dogs to exist on public grasslands, and to allow translocations of prairie dogs from urban areas where development destroys prairie dog habitat. People's knowledge of prairie dogs is directly related to their personal experience. Rather than spend money on poisoning prairie dogs, agency land managers can spend money on developing watchable wildlife sites at prairie dog colonies. Because prairie dogs are active in the daytime and so many other animals are associated with prairie dog towns, watchable wildlife sites on prairie dog towns would be an excellent draw for tourists, and would provide first-hand experience with prairie dogs to a larger segment of the general public. At least two states, Montana and Kansas, already have such entities and they are popular (R. Reading, personal communication). Modest fees can be charged for maintenance and upkeep of the watchable wildlife sites.

As a way of reaching a broader segment of the general public, conservation organizations can raise funds to support the production of television shows, magazine articles, and conferences on the life of prairie dogs. Many people today get a considerable amount of information from television, which can reach a broad segment of the general public.

Decisions about poisoning or shooting prairie dogs are made by politicians and by agency land managers who are responsive to their constituencies. So far, the constituencies with negative views have dominated much of the political and agency processes, resulting in prairie dogs moving toward extinction. But views can change. Positive signs of change include the rising numbers of volunteer organizations (such as the Rocky Mountain Animal Defense and the Prairie Dog Coalition), and conferences such as the Prairie Dog Summits of 2001 and 2003 in Denver, and Prairie Dog Technical Conferences in Fort Collins, Colorado. As more people realize that prairie dogs have a critical role in the ecosystem and have a complex set of abilities, more people will be able to identify with prairie dogs and find them non-threatening.

9

Room to Hope

Prairie dogs are important, exciting mammals. Within their small, tawny bodies are key traits that link together an entire prairie ecosystem and carry irreplaceable and unique lessons about animal communication, sociality, and the human–animal relationship. Learning about prairie dogs, their plight, and their contributions to natural ecosystems does more than give us a sense of desperation at their current decline. It also gives us respect for the value of their lives, and a more informed platform from which to make decisions about how we should interact with them.

Our research indicates that prairie dogs may be one of the most complex social communicators documented so far in the vertebrate world. Not only can they communicate the presence of different species of predators but also different individuals within the same species (see chapter 4). Prairie dog alarm calls are given in response to a predator and accompanied by different evasive behaviors. In that sense, they provide a Rosetta Stone for decoding the alarm call communication system. We can see the predator approaching, we can record the alarm calls, and we can document the escape behaviors on video for further analysis in the laboratory. We can also experimentally manipulate the communication system by introducing objects that produce alarm calls, and study the characteristics of those objects and how those characteristics are reflected in the acoustic structure of the calls. For this reason the calls of prairie dogs are much easier to interpret than other species, such as monkeys and whales, that are expected to have sophisticated communication systems. Prairie dogs provide us with a straightforward communication system that is easily verifiable through experimentation and has the potential to contribute greatly to furthering our knowledge about how animals communicate. Also, prairie dogs have different alarm call dialects between colonies. Colony dialects seem to vary, not only with distance from one colony to another, but also with differences between the vegetation structures of colonies, through which the calls travel.

Studying prairie dogs provides the opportunity for systematically exploring the potential ability of animals to communicate, and also the evolutionary mechanisms that lead to communication complexity. Our studies suggest that prairie dogs have a number of characteristics in their communication system that are proposed by linguists as reflecting language, such as semantics, productivity, displacement, and duality. This leads to the question, How many other species of animals might have similar characteristics in their communication systems, if we did the right experiments and looked in the right places? Answering the above question has the potential to redefine current notions of the intellectual capacity of non-human animals.

The social complexity and variety of social pairings exhibited on prairie dog towns also provide the potential to explore questions about what drives sociality in animals and how environment and resources may affect social group composition (see chapter 3). In Gunnison's prairie dogs, the structure of the social group is related to the food resources on a territory (Slobodchikoff 1984; Travis and Slobodchikoff 1993; Verdolin 2007). A model for how this kind of flexibility of social structure can arise suggests that when resources are uniformly distributed, prairie dogs will form small groups of either male and female pairs, or single animals, while when resources are patchily distributed, the prairie dogs will form larger groups consisting of either single males and multiple females, or multiple males and multiple females (Slobodchikoff 1984). This kind of flexible social structure in response to resource distributions might be directly comparable to the kinds of distributions of social groups found among humans.

Understanding the relationship and importance of social structures to species survival is not only an interesting academic question, it has been used in conservation efforts to increase the chance of success. For example, during release of wolves into Yellowstone, maintaining their social groups intact increases wolf survival in a new location and also helps decrease the stress of relocation and the tendency for the wolves to attempt to run back to their home lands (Bangs and Fritz 1996). In addition, understanding the drive of young condors to socialize and bond with their food providers allows researchers and conservationists to use techniques that decrease the chance for condors to bond with humans, thus increasing their independence when they are released into the wild (Cohn 1999). In Vermejo Park Ranch, New Mexico, maintaining intact social groups when relocating prairie dogs results in lower mortality levels and higher reproductive rates (Shier 2006). Therefore, learning more about the importance and characteristics of prairie dog social systems potentially carries applicable lessons, and practical benefits for relocation and conservation programs in a variety of species.

Prairie dog colonies are also model systems to explore questions of evolutionary biology and genetics. Because prairie dog towns exist in many different sizes and degrees of isolation, there is ample opportunity to explore evolutionary questions about the effects of isolation on gene pools. The results of such studies can be applied to major questions in conservation biology about the effects of habitat loss and isolation on species genetic variability and fitness. In addition, combining knowledge of social behavior and genetics can offer insight into how social behaviors may interact with environmental factors to influence the degree of inbreeding and genetic variability in populations. Thus, prairie dogs provide a convenient subject for many evolutionary, social, and communication studies that contribute to theoretical scientific knowledge as well as applied conservation challenges.

Prairie dogs are also directly important to sustaining healthy grasslands. They are generally accepted as the keystone species of American grasslands (see chapter 6). By digging burrow systems, prairie dogs affect soil nutrient cycling, plant and animal diversity, and small mammal abundance (O'Meilia et al. 1982; Agnew et al. 1986; Whicker and Detling 1988). More than 160 vertebrate species are either casually or intimately associated with prairie dog towns, relying on prairie dogs for habitat and food resources (Clark et al.1982; Kotliar et al. 1999). The widespread reduction of prairie dogs has had a domino effect on a host of species, some of which are threatened or endangered. Several species have had a corresponding decline alongside prairie dogs, including the swift fox, the burrowing owl, the golden eagle, the ferruginous hawk, the mountain plover, and the black-footed ferret (*Vulpes velox,* Knowles and Knowles 1994; *Athene cunicularia,* Knowles and Knowles 1994; Desmond et al. 2000; *Aquila chrysaetos,* Cully 1991; *Buteo regalis,* Knowles and Knowles 1994; Miller et al. 1994; *Charadrius montanus,* Knowles and Knowles 1994; *Mustela nigripes,* Clark 1989). The extinction of prairie dogs would likely cause the extinction of other species as well.

Techniques, insights, and solutions developed to conserve prairie dogs can also contribute to our understanding of how to counteract the larger trend of global biodiversity decline. Prairie dog decline is not an isolated phenomenon. Almost 20% of vertebrates are currently in danger of extinction. This extinction is human caused. The world has lost almost 650 species in the last 400 years (Chapin et al. 2000). This rate is more than three orders of magnitude higher than background, or non-human caused, extinction levels (Nott et al. 1996). Considering not just vertebrates but all animals combined, we currently have about one million species that are known to science, and perhaps an estimated 20–30 million that have not yet been found (Wilson 1992). Estimates have suggested that we lose about four species per hour, or about 100 species per day, or 36,500 per year (WRI 1998). We are responsible for the biggest number of extinctions since the dinosaurs died out some 65 million years ago (Bradford and Dorfman 2002).

Prairie dogs are subject to the same forces of human-caused extinction as other vertebrates. The top 10 reasons for extinction of vertebrates in the United States are as follows: 1) interactions with non-native species; 2) urbanization; 3) agriculture; 4) outdoor recreation and tourism; 5) ranching; 6) water diver-

Table 9.1.   Relating major causes of extinction in vertebrates to prairie dog decline.

Top 10 causes of species extinction (Czech and Krausman 1997)	Prairie dogs	Direction of threat
1. Interaction with non-native species	Plague	Unknown
2. Urbanization	Farm/ranchland to sub-urban/urban conversion	Increasing (Heimlich and Anderson 2001)
3. Agriculture	Poisoning, crops not habitat	Decreasing in west— converting to suburban (Riebsane et al. 1996)
4. Outdoor recreation and tourism	Recreational shooting	Unknown
5. Ranching	Poisoning, shooting	Decreasing in west (Sullins et al. 2002)— ranching converted to ex-urban suburban
6. Water diversions	Not likely	
7. Modified fire regimes	Not likely	
8. Pollution of water, air, soil	Possible, not documented	Locally dependent trends
9. Energy exploration	Possible, not documented	Increasing in west (Bryner 2003)
10. Industrial and military activity	Possible, not documented	Unknown

sions; 7) modified fire regimes; 8) pollution of water, air, and soil; 9) energy exploration; and 10) industrial and military activity (Czech and Krausman 1997). Prairie dogs are affected by some of these forces (Table 9.1).

Even though each case of species extinction is unique, prairie dog decline can be evaluated as it exists within the larger picture of human land-use in the United States. For example, the top five causes of vertebrate extinction in the United States apply directly to prairie dogs. Thus, the decline of prairie dogs is not unique but rather quite typical of what a variety of other vertebrate species are experiencing. This signifies that prairie dog decline is part of a larger pattern driven by human choices about land-use, and that a systematic effort to reduce these threats would be valuable in stopping the decline, not only of the prairie dog, but of other species as well. In addition, the fact that drivers of prairie dog decline are also the top five causes of the major extinction event for all vertebrates means that lessons learned in reversing or stopping drivers of prairie dog decline could be applicable, with some variations, to other species.

Just as prairie dogs are not isolated in the causes of decline, the consequences of prairie dog decline extend beyond the loss of prairie dogs. Prairie dogs provide a variety of values for human and non-human animals that would be lost with their extinction. However, in documenting the magnitude and extent of this loss, we find one possible key to saving this incredible genus. The potential enormity of ecosystem services provided by native grasslands (including prairie dog habitats) in terms of pure economic benefit alone could be a powerful tool to wield in the progress toward prairie dog conservation (see chapter 6). Table 9.2 summarizes information on possible consequences of prairie dog decline (also see chapters 6 and 7).

While once covering the North American prairie for as far as the eye could see, all five species of prairie dogs have declined precipitously (see chapter 8). Black-tailed prairie dogs have declined as much in range as grizzlies and wolves, both of which are listed as endangered species and have recovery plans in place. Yet prairie dogs lack the backing of strong, affirmative public opinion. It has been exceedingly difficult for proponents to get endangered status approved for the black-tailed prairie dog and there is no federal protection in place for any prairie dog species except the Utah prairie dog, which is listed as threatened, and the Mexican prairie dog, which is listed as endangered under CITES and the ESA (see chapter 7).

Table 9.2. Possible consequences of prairie dog decline: summary table with references to appropriate chapter in book.

Consequence	Explanation	Chapter reference
Highly dependent species decline or become extinct	Black-footed ferret, swift fox, burrowing owl, golden eagle, ferruginous hawk have all experienced a corresponding decline with prairie dogs	Chapters 6 and 7
Associated species decline	Possibly up to 160 different species that use prairie dog towns could be adversely affected by their absence	Chapters 6 and 7
Groundwater decline	Prairie dog burrowing may increase percolation into aquifers. If prairie dog towns are converted to ranchland and suburban or urban development, soil is compacted or paved, which could increase run-off and decrease acquifer recharge.	Chapter 6
Inhibited soil formation/ aeration processes	Prairie dogs contribute to soil aeration and nutrient cycling	Chapters 6 and 7
Decrease in native herbaceous plants		Chapters 6 and 7
Tree encroachment into grasslands	Prairie dog selected grazing and granivory maintains grasslands	Chapter 6
Loss or increase of vegetation productivity	Prairie dog grazing alters productivity of grasslands	Chapters 6 and 7
Lost opportunity for theoretical and practical scientific research	Research on prairie dogs currently contributes important discoveries to the fields of: animal communication, animal behavior, evolutionary biology, genetics, ecology	Chapters 3, 4, and 5
Recreation	Loss of a variety of activities that occur on prairie dog towns including: watchable wildlife, hunting, plant/herb collecting, hiking, birding	Chapter 7
Heritage/history	Prairie dogs are a unique part of American history and natural heritage, unique to North America and mentioned in historical references	Chapters 5, 7, and 8

However, we would not have spent such effort compiling this information and writing this book if we did not believe that there was a chance to turn the trajectory of decline around. The desperation of this situation is not new. Rather, it is often the case for environmental problems. Many people in many other situations have experienced the feeling of being overwhelmed by the magnitude and complexity of conservation problems, as Allen Hammond (1998) points out in the starting pages of his book, *Which World? Scenarios for the 21st century*. Hammond, a senior scientist in strategic analysis of global trends at the World Resources Institute, says that his students seem to believe that everything in the world is going to get worse, while he, a person who sees seemingly desperate and disparaging trends every work day, believes that the future is ultimately more hopeful than the trends suggest.

Hammond goes on to speak about two dangers in peoples' response to environmental problems—simplistic optimism, and simplistic pessimism. And while he states that simplistic optimism is potentially more dangerous and more widespread than simplistic pessimism, he argues that both lead to enormous dysfunction in terms of inhibiting productive solutions to world environmental problems. To combat both of these dangers in our situation, it is essential to delve deep into the trends of the prairie dog dilemma and attempt to find the root causes of the conservation challenge that we face. Hopefully, the information in this book can serve as a starting point for a deeper level of focus and reflection on the prairie dog conservation challenge.

Stories like those about prairie dog decline are likely to continue unless human societies can muster the knowledge, understanding, compassion, and will to make fundamental and widespread changes in our relationship with the non-human world. An expansive redefinition of the human relationship to nature requires not only an open international, national, state, county, and community-wide dialog, it also requires each human being, in their own way, to reexamine and redefine their physical and emotional relationship to non-human entities. Only through a multi-level discourse and examination of this human–nature relationship can we expect change to occur.

In the March 1993 issue of *Conservation Biology*, James Karr wrote a letter about advocacy and responsibility and how these two words have influenced, and many times inhibited, the involvement of conservation biologists in contributing to environmental decision-making. He debunks the notion that scientists who speak out about natural systems are advocates rather than responsible professionals. In his letter, Karr makes the analogy between an engineer who knows that a bridge is structurally flawed and an ecologist who knows that an ecosystem is being destroyed, saying that in both cases, it is a professional obligation to speak out to try to correct the problem.

Key in this assertion is that scientists use real knowledge and scientific expertise to give the most conscientious advice that they can. Often the system dynamics and feedbacks are so varied and complex that it is difficult to tell with absolute certainty what the ecological and social outcomes of our actions will be. Uncertainty is often what leads scientists to hesitate in sharing their knowledge, and often keeps politicians and land managers from feeling comfortable about acting. However, our understanding of the deleterious effects that humans have on non-human life forms and systems is more sophisticated now than ever before, and more and more professional scientists are coming to the conclusion that, in the face of uncertainty, the risks of inaction far outweigh the risks of acting to conserve nature (Weeks 1997; see also literature on the precautionary approach, Gilligan 2003; Foster et al. 2000).

In terms of prairie dogs, the risks of inaction are multiple and compounded by the uniqueness of this genus. Prairie dogs have evolved specifically within North American grasslands in conjunction with other grassland species—they do not exist on any other continent. All five species of prairie dogs are irreplaceable, and like many other species existing in the periphery of our social value system, their future is uncertain.

Thus, not only professional scientists, but land managers, politicians, concerned citizens, and educators have a responsibility to reexamine our relationship with prairie dogs. Just as scientists have the responsibility to pass on their knowledge of the consequences of human actions on living systems, land managers and politicians have the responsibility to take that knowledge and use their expertise to develop political and management strategies that limit human-caused destruction of the natural environment. Concerned citizens have the responsibility to support those land managers and politicians who create positive change and to recall those that do not. We now know something about the "broken bridges" in our native grassland ecosystems, and we know something about alternative ways of dealing with current conservation challenges. Thus, we have the responsibility to put this knowledge on the table and to start acting on it. We hope that in reading these pages everyone can come away with a better understanding of the risks inherent in prairie dog decline, what prairie dog conservation entails, and how conservation of prairie dogs fits into the larger challenge of conservation on multiple scales, from species, to ecosystems, to global systems.

Our relationship with prairie dogs started through our mutual interests in animal communication. At the outset we were focused on decoding prairie dog alarm calls. But, in addition to the bark of the prairie dog, our research tapes are full of the sounds of the other animals that use prairie dog towns: elk bugling, coyotes howling, deer stomping, rattlesnakes buzzing, hawks screeching, owls hooting, and the constant drone of a multiplicity of insects. Listening to these tapes, trying to separate out the alarm calls of focus, reminded us how important prairie dog colonies are to other grassland species. Prairie dog alarm calls are not bone-chillingly beautiful like a wolf howl, or sweet and melodic like the meadowlark. Prairie dog calls are harsh, and short, and raspy. Yet a tremendous loss to the prairie, and to American society, is represented in their potential silence.

Through our work we also came into contact with many human voices, varied in tone, age, and upbringing: voices of admiration and surprise toward prairie dogs, voices of hate and condemnation, voices of apathy, and voices of concern. No voice matched any one face or walk of life. It surprised us just as much to hear how humans spoke about prairie dogs as it was for us to study how prairie dogs communicated with each other. And we were reminded how much the fate of the prairie dog is in human hands.

Appendix:
Where to See Prairie Dogs

Whether it is in a natural environment or in a zoological park, prairie dogs draw hundreds of thousands of visitors each year. Prairie dogs can be seen in cities and towns throughout their range. More progressive cities and towns have developed watchable wildlife sites (e.g. Flagstaff, Arizona). Others have plans in the making (e.g. Santa Fe, New Mexico). While not an exhaustive list, below are some of the National Monuments, Wildlife Refuges and Parks that are home to prairie dogs:

Arapaho National Wildlife Refuge, CO
Badlands National Park, SD
Bryce Canyon National Park, UT
Buffalo Lake National Wildlife Refuge, TX
Charles M. Russell National Wildlife Refuge, MT
Conata Basin in Buffalo Gap National Grassland, SD
Devils Tower National Monument, WY
Dixie National Forest, UT
Fort Niobrara National Wildlife Refuge, NE
Grasslands National Park, Saskatchewan Canada
Halfbreed Lake National Wildlife Refuge, MT
Kirwin National Wildlife Refuge, NE
Lacreek National Wildlife Refuge, SD
Lake Mason National Wildlife Refuge, MT
Maxwell National Wildlife Refuge, NM

Ouray National Wildlife Refuge, UT
Petrified Forest National Park, AZ
Sullys Hill National Game Preserve and Wildlife Refuge, ND
Theodore Roosevelt National Park, ND
Ul Bend National Wildlife Refuge, MT
War Horse National Wildlife Refuge, MT
Wind Cave National Park, SD
Witchita Mountains Wildlife Refuge, OK

If you do not live near these areas or are not planning a trip out west, zoos worldwide are hosts to prairie dogs. Black-tailed prairie dogs are almost exclusively the species that you will find in zoos. Every attempt has been made to compile a complete list for accredited zoos in the United States. In addition, a shorter list of international zoos is included. If a location near you is not listed, call your local zoo to find out if they have prairie dogs. Typically, prairie dog exhibits are located in the Children's Zoo of most parks. The benefit to the visitor is an up close view of prairie dogs, usually including a series of viewing bubbles right in the exhibit.

Arizona

Arizona-Sonora Desert Museum
2021 North Kinney Road
Tucson, Arizona 85743

The Phoenix Zoo
455 North Galvin Parkway
Phoenix, Arizona 85008
Tel: (602)273-1341
Fax: (602)273-7078

Arkansas

Little Rock Zoo
1 Jonesboro Drive
Little Rock, Arkansas 72205-5401
Tel: (501)666-2406
Fax: (501)666-7040

California

Fresno Chaffee Zoo
894 W. Belmont Avenue
Fresno, California 93728

San Francisco Zoo
1 Zoo Road
San Francisco, California 94132
Crawl-in burrows let kids experience the animals' underground homes; and a cutaway model of a prairie dog den shows how the burrows contain nursery, nesting chamber, and latrine rooms. Visitors can help feed the animals at scheduled times.

Colorado

The Cheyenne Mountain Zoo
4250 Cheyenne Mountain Zoo Road
Colorado Springs, Colorado 80906-5755
Tel: (719)633-9925
Fax: (719)633-2254

Connecticut

Connecticut's Beardsley Zoo
1875 Noble Avenue
Bridgeport, Connecticut 06610

Florida

Caribbean Gardens
1590 Goodlette Road
Naples, Florida
Tel: (239)262-5409

Palm Beach Zoo at Dreyer Park
1301 Summit Boulevard
West Palm Beach, FL 33405-3035
Tel: (561)533-0887
Fax: (561)585-6085

Indiana

Fort Wayne Zoo
3411 Sherman Boulevard
Fort Wayne, Indiana 46808
Children's Zoo

Mesker Park Zoo and Botanic Garden
2421 Bement Avenue
Evansville, Indiana 47720-8206
Tel: (812)435-6143
Fax: (812)435-6140

Iowa

Blank Park Zoo
7401 SW 9th Street
Des Moines, Iowa 50315-6667
Tel: (515)285-4722
Fax: (515)285-1487

Kansas

Hutchinson Zoo
PO Box 1567
6 Emerson Loop East
Hutchinson, Kansas 67504
Children can view the prairie dogs through a specially designed tunnel to safely get up close to these timid creatures.

Rolling Hills Wildlife Adventure
625 N Hedville Road
Salina, Kansas 67401-9764
Tel: (785)827-9488
Fax: (785)827-3738

Sedgwick County Zoo
5555 W Zoo Boulevard
Wichita, Kansas 67212-1698
Tel: (316)266-8201
Fax: (316)942-3781

Sunset Zoo
2333 Oak Drive
Manhattan, KS 66502-3824
Tel: (785)587-2737
Fax: (785)587-2730

Maryland

Salisbury Zoo
PO Box 2979
755 South Park Drive
Salisbury, Maryland 21802-2979
Tel: (410)548-3188
Fax: (410)860-0919

Michigan

The Detroit Zoo
PO Box 39
8450 West Ten Mile Road
Royal Oak, Michigan 48068-0039
Tel: (248)398-0900

Minnesota

Lake Superior Zoo
7210 Fremont Street
Duluth, Minnesota 55807

Minnesota Zoo
13000 Zoo Boulevard
Apple Valley, Minnesota 55124-8199
Tel: (952)431-9200
Fax: (952)431-9300

Mississippi

Jackson Zoological Park
2918 W Capitol Street
Jackson, Missippi 39209-4293
Tel: (601)352-2581
Fax: (601)352-2594
Located in the Discovery Zoo area.

New York

Bronx Zoo
2300 Southern Boulevard
Bronx, New York 10460-1090
Tel: (718)220-5100
Fax: (718)733-7748
Seasonal exhibit in the Children's Zoo.

Prospect Park Zoo
450 Flatbush Avenue
Brooklyn, New York 11225
The Prairie Dog habitat is located on the Zoo's Discovery Trail, a 2.5-acre winding path comprised of outdoor animal exhibits that allow visitors to experience the animals' habitat firsthand. The interactive prairie dog town even includes child-sized tunnels, so that kids and prairie dogs can meet nose-to-nose.

Ross Park Zoo
60 Morgan Road
Binghamton, New York 13903
Tel: (607)724-5461

Queens Zoo
Flushing Meadows Corona Park
Queens, New York

North Dakota

Dakota Zoo
Arbor Avenue
Bismarck, North Dakota

Red River Zoo
4220 21st Avenue SW
Fargo, North Dakota 58104-8603
Tel: (701)277-9240
Fax: (701)277-9238

Ohio

Columbus Zoo and Aquarium
PO Box 400
9990 Riverside Drive
Powell, Ohio 43065
Prairie dog colony in the North America region at the Columbus Zoo and Aquarium.

Pennsylvania

Elmwood Park Zoo
1661 Harding Boulevard
Norristown, Pennsylvania 19401
Tel: (610)277-3825

Lehigh Valley Zoo
PO Box 519
5150 Game Preserve Road,
Schnecksville, Pennsylvania 18078-0519
Tel: (610)799-4171
Fax: (610)799-4170

Philadelphia Zoo
3400 W Girard Avenue
Philadelphia, Pennsylvania 19104
Tel: (215)243-1100
Fax: (215)243-5385

South Dakota

Bramble Park Zoo
Bramble Park
Watertown, SD
Tel: (605)882-6269

Tennessee

Chattanooga Zoo
1101 McCallie Avenue
Chattanooga, Tennessee 37404
Tel: (423)697-1322
Fax: (423)697-1329

Knoxville Zoo
PO Box 6040
3500 Knoxville Zoo Drive
Knoxville, TN 37914

Texas

Abilene Zoological Park
2070 Zoo Lane
Abilene, TX 79602-1996
Tel: (325)676-6085
Fax: (325)676-6084

El Paso Zoo
4001 E. Paisano
El Paso, Texas 79905-4223
Tel: (915)521-1850
Prairie dogs located in the Highlights of the Americas exhibit.

Fort Worth Zoo
1989 Colonial Parkway
Fort Worth, Texas 76110
Tel: (817)759-7555
Fax: (817)759-7501
The new Prairie Dog Exhibit opened in the Texas Wild! section of the Fort Worth Zoo in 2001. There are several prairie dogs on display and they share their exhibit with burrowing owls.

Houston Zoo
1513 N Macgregor Drive
Houston, Texas 77030-1603
Tel: (713)533-6801
Fax: (713)533-6802

San Antonio Zoo
3903 North St. Mary's Street,
San Antonio, Texas 78212-3199
The new prairie dog exhibit at Kronkosky's Tiny Tot Nature Spot.

Virginia

Mill Mountain Zoo
Roanoke, Virginia 24034-3484
Tel: (540)343-3241
Fax: (540)343-8111

Virginia Zoological Park
3500 Granby Street
Norfolk, Virginia 23504-1329
Tel: (757)441-2374
Fax: (757)441-5408

Washington, DC

National Zoo
3001 Connecticut Avenue NW
Washington, DC 20008.

Wisconsin

Henry Vilas Zoo
702 South Randall Avenue
Madison, Wisconsin 53715-1665
Planned exhibit. Contact for details.

New Zoo
4378 Reforestation Road
Green Bay, Wisconsin 54313

Washington

Woodland Park Zoo
601 N 59th Street
Seattle, Washington 98103-5897
Tel: (206)684-4880
Fax: (206)684-4854

International

Austria
Tiergarten Schönbrunn Zoo Vienna
Schönbrunner Tiergarten Ges.m.b.H.
Maxingstrasse 13 b, A-1130 Wien

Bahamas
Ardastra Gardens Zoo and Conservation Center
Nassau

Belgium
Olmense Zoo
Bukenberg 45
B-2491 Olmen-Balen
Tel: +32-(0)14-30.98.82
Fax: +32-(0)14-30.23.46

Canada
Assiniboine Park Zoo
2355 Corydon Avenue
Winnipeg, Manitoba R3P 0R5
Tel: +1 204 986 6921

Calgary Zoo
1300 Zoo Road NE
Calgary, Alberta T2E 7V6
Tel: (800)588-9993

Fort Whyte Centre
1961 McCreary Road
Winnipeg, Manitoba R3P 2K9
Tel: +1 204 989 8335

Saskatoon Forestry Farm Park and Zoo
1903 Forestry Farm Park Drive
Saskatoon, Saskatchewan S7S 1G9
Tel: (306)975-3382

Toronto Zoo
361A Old Finch Avenue
Scarborough, Ontario M1B 5K7
Tel: (416)392-5901
Fax: (416)392-5934

Valley Zoo
Buena Vista Road and 134 Street
Edmonton, Alberta
Tel: (780)496-8787

Czech Republic
Zoo Děčín
Zizkova 15, CZ-405 02 Děčín IV
Tel/Fax: 000420 412 531 164
Email: info@zoodecin.cz
Internet: www.zoodecin.cz

Germany
Zoo Köln
Riehler Straße 173
50735 Köln
Tel: 02 21/77 85 0
Fax: 02 21/77 85 111
Email: info@zoo-koeln.de
Internet: www.zoo-koeln.de

Greece
Attica Zoological Park
Athens

Japan
Shizuoka Municipal Nihondaira Zoo
Shizuoka

Ueno Zoological Gardens
Ueno Park, Taito-ku, Tokyo

United Kingdom
Banham Zoo
The Grove
Banham, Norfolk, NR16 2HE
Tel: 01953 887771
Fax: 01953 887445

Chester Zoo
Upton-by-Chester
Chester
CH2 1LH
Tel: 01244 380 280
Fax: 01244 371 273

Folly Farm Family Adventure Park
Begelly
Kilgetty
Pembrokeshire
SA68 OXA

Twycross Zoo
Atherstone
Warwickshire
CV9 3PX
Tel: 01827 880250/880440
Fax: 01827 880700

Whipsnade Wild Animal Park
Dunstable
Bedfordshire LU6 2LF
Tel: 01582 872171
Fax: 01582 872649

Bibliography

Ackers, S. H. 1997. The communicative function of variation in Gunnison's prairie dog *(Cynomys gunnisoni)* alarm calls. PhD diss., Northern Arizona University.

Ackers, S. H., and C. N. Slobodchikoff. 1999. Communication of stimulus size and shape in alarm calls of Gunnison's prairie dogs. *Ethology* 105:149–62.

Adams, R. A., B. J. Lengas, and M. Bekoff. 1987. Variations in avoidance responses to humans by black-tailed prairie dogs *(Cynomys ludovicianus)*. *Journal of Mammalogy* 68:686–89.

Agnew, W., D. W. Uresk, and R. M. Hansen. 1986. Flora and fauna associated with prairie dog colonies and adjacent ungrazed mixed-grass prairie on South Dakota. *Journal of Range Management* 39:135–39.

Allison, P. S., A. W. Leary, and M. J. Bechard. 1995. Observations of wintering ferruginous hawks *(Buteo regalis)* feeding on prairie dogs *(Cynomys ludovicianus)* in the Texas panhandle. *Texas Journal of Science* 47:235–37.

Andelt, W. F. 2006. Methods and Economics of Managing Prairie Dogs. In *Conservation of Black-tailed Prairie Dogs*, ed. J. L. Hoogland, 129–38. Washington, DC: Island Press.

Anderson, E., S. C. Forrest, T. W. Clark, and L. Richardson. 1986. Paleo-biology, biogeography, and systematics of the black-footed ferret, *Mustela nigripes*. In Audubon and Bachman 1851, *Great Basin Naturalist Memoirs* 8:11–26.

Anderson, S. H., and E. S. Williams. 1997. Plague in a complex of white-tailed prairie dogs and associated small mammals in Wyoming. *Journal of Wildlife Diseases* 33:720–32.

Anthony, A. 1955. Behavior patterns in a laboratory colony of prairie dogs, *Cynomys ludovicianus*. *Journal of Mammalogy* 36:69–78.

Archer, S., and J. K. Detling. 1986. Evaluation of potential herbivore mediation of plant water status in a North American mixed-grass prairie. *Oikos* 47:287–91.

Archer, S., M. G. Garrett, and J. K. Detling. 1987. Rates of vegetation change associated with prairie dog *(Cynomys ludovicianus)* grazing in North American mixed-grass. *Vegetatio* 72:159–66.

Arjo, W. M. 1992. Agonistic behaviors and dominance hierarchies in Gunnison's prairie dogs *(Cynomys gunnisoni)*. MS thesis, Flagstaff: Northern Arizona University.

Arrowood, P. C., C. A. Finley, and B. C. Thompson. 2001. Analyses of burrowing owl populations in New Mexico. *Journal of Raptor Research* 35:362–70.

Axelrod, D. I. 1985. Rise of the Grassland Biome, Central North America. *Botanical Review* 51:163–201.

Bak, J. M., K. G. Boykin, B. C. Thompson, and D. L. Daniel. 2001. Distribution of wintering ferruginous hawks *(Buteo regalis)* in relation to black-tailed prairie dog *(Cyomys ludovicianus)* colonies in southern New Mexico and northern Chihuahua. *Journal of Raptor Research* 35:142–29.

Baker, H. 1956. *Mammals of Coahuila, Mexico.* Lawrence: University of Kansas Museum of Natural History.

Bakko, E. B., and L. N. Brown. 1967. Breeding biology of the white-tailed prairie dog, *Cynomys leucurus,* in Wyoming. *Journal of Mammalogy* 48:100–12.

Bakko, E. B., and J. Nahorniak. 1986. Torpor patterns in captive white-tailed prairie dogs *(Cynomys leucurus)*. *Journal of Mammalogy* 67:576–78.

Bakko, E. B., W. P. Porter, and B. A. Wunder. 1988. Body temperature patterns in black-tailed prairie dogs in the field. *Canadian Journal of Zoology* 66:1783–89.

Bangert, R., and C. N. Slobodchikoff. 2000. The Gunnison's prairie dog structures high desert grassland landscape as a keystone engineer. *Journal of Arid Environments* 46:357–69.

Bangert, R. K., and C. N. Slobodchikoff. 2004. Prairie dog engineering indirectly affects beetle movement behavior. *Journal of Arid Environments* 56:83–94.

———. 2006. Conservation of prairie dog ecosystem engineering may support beta and gamma diversity. *Journal of Arid Environments* 67:100–15.

Bangs, E. E., and S. H. Fritz. 1996. Reintroducing the gray wolf to central Idaho and Yellowstone National Park. *Wildlife Society Bulletin* 24:402–13.

Barko, V. A., J. H. Shaw, and D. M. J. Leslie. 1999. Birds associated with black-tailed prairie dog colonies in southern short-grass prairie. *Southwestern Naturalist* 44:484–89.

Barnes, A. M. 1982. Surveillance and control of bubonic plague in the United States. *Symposia of the Zoological Society of London* 50:237–70.

Beard, M. L., S. T. Rose, A. M. Barnes, and J. A. Montenieri. 1992. Control of *Oropsylla hirsute,* a plague vector, by treatment of prairie dog burrows with 0.5% permetrin dust. *Journal of Medical Entomology* 29:25–29.

Bell, W. R. 1921. *See* U.S. Department of Agriculture.

Berger, J. 1978. Group size, foraging, and anti-predator ploys: An analysis of bighorn sheep decisions. *Behavioural Ecology and Sociobiology* 4:91–99.

Biggins, D. E., and M. Y. Kostoy. 2001. Influences of introduced plague on North American mammals: Implications from ecology of plague in Asia. *Journal of Mammalogy* 82:906–16.

Biggins, D. E., B. J. Miller, L. Hanebury, R. Oakleaf, A. Farmer, R. Crete, and A. Dood. 1993. A technique for evaluating black-footed ferret habitat. In *Proceedings of the symposium on the management of prairie dog complexes for reintroduction of the black-footed ferret,* ed. J. Oldenmeyer, U.S. Department of the Interior, Fish and Wildlife Services.

Biggins, D. E., J. G. Sidle, D. B. Seery, and A. E. Ernst. 2006. Estimating the abundance of prairie dogs. In *Conservation of the black-tailed prairie dog,* ed. J. L. Hoogland, 94–107. Washington, DC: Island Press.

Bock, C. E., and J. H. Bock. 1993. Cover of perennial grasses in southeastern Arizona in relation to livestock grazing. *Conservation Biology* 7:371–77.

Bonham, C. D., and A. Lerwick. 1976. Vegetation changes induced by prairie dogs on shortgrass range. *Journal of Range Management* 29:221–25.

Bradford, L., and A. Dorfman. 2002. The state of the planet. *Time Magazine,* August 26.

Bragg, T., and A. A. Steuter. 1996. Prairie ecology—the mixed prairie. In *Prairie Conservation: Preserving North America's most endangered ecosystem,* ed. F. Sampson and F. Knopf. Washington, DC: Island Press.

Brizuela, M. A., J. K. Detling, and M. Cid. 1986. Silicon concentration of grasses growing in sites with different grazing histories. *Ecology* 67:1098–1101.

Bronson, M. T. 1979. Altitudinal variation in the life history of the golden-mantled ground squirrel *(Spermophilus lateralis). Ecology* 60:272–79.

Brown, D. E. 1982. *Desert plants: Biotic communities of the American Southwest and Mexico.* Tucson: University of Arizona Press.

Brown, D. E., C. H. Lowe, and C. P. Pase. 1980. A digitized systematic classification system for the biotic communities of North America with community (series) and association examples for the Southwest. *General Technical Report RM-73.* USDA Forest Service Rocky Mountain Forest and Range Experimental Station, Fort Collins, CO.

Brown, J. H. 1978. The theory of insular biogeography and the distribution of birds and mammals. *Great Basin Naturalist Memoirs* 2:209–27.

Brown, M. T., and R. A. Herendeen. 1996. Embodied energy analysis and EMERGY analysis: A comparative view. *Ecological Economics* 19:219–35.

Brown, M. T., and S. Ulgiati. 1997. Emergy-based indices and ratios to evaluate sustainability: Monitoring economies and technology toward environmentally sound innovation. *Ecological Engineering* 9:51–69.

Bryner, G. C. 2003. Coalbed methane development: the costs and benefits of an emerging energy resource. *Natural Resources Journal* 43:519–39.

Bureau of Sport Fisheries and Wildlife. 1968. Rare and endangered fish and wildlife of the United States. Washington, DC: U.S. Government Printing Office.

Burnett, S., and B. Allen. 1998. Endangered Species Economics. *Brief Analysis* No. 276. Dallas, TX: National Center for Policy Analysis.

Burns, J. A., D. L. Flath, and T. W. Clark. 1989. On the structure and function of white-tailed prairie dog burrows. *Great Basin Naturalist* 49:517–24.

Burroughs, A. L. 1947. Sylvatic plague studies: The vector efficiency of nine species of fleas compared with *Xenopsylla cheopis*. *Journal of Hygiene* 45:371–96.

Cable, K. A., and R. M. Timm. 1987. Efficacy of deferred grazing in reducing prairie dog reinfestation rates. In *Eighth Great Plains Wildlife Damage Control Workshop Proceedings, 28–30 April 1987,* 46–49. Rapid City, SD.

Cables, H. S. 1993. Acting Director, USNPS to David Roemer, 19 May 1993.

Campbell, T. M. III, and T. W. Clark. 1981. Colony characteristics and vertebrate associations of white-tailed and black-tailed prairie dogs in Wyoming. *American Midland Naturalist* 105:268–76.

Caraco, T. 1979. Time budgeting and group size: A test of theory. *Ecology* 60:618–27.

Carlson, D. C., and E. M. White. 1987. Effects of prairie dogs on mound soils. *Soil Science Society of America Journal* 51:389–93.

Cartron, J.-L. E., P. J. Polechla, Jr., and R. R. Cook. 2004. Prey of nesting ferruginous hawks in New Mexico. *Southwestern Naturalist* 49:270–76.

Ceballos, G., J. Pacheco, and R. List. 1999. Influence of prairie dogs *(Cynomys ludovicianus)* on habitat heterogeneity and mammalian diversity in Mexico. *Journal of Arid Environments* 41:161–72.

Chapin, F. S., E. S. Zavaleta, V. T. Eviner, R. L. Naylor, P. M. Vitousek, H. L. Reynolds, D. U. Hooper, S. Lavorel, O. E. Sala, S. E. Hobbie, M. C. Mack, and S. Diaz. 2000. Consequences of changing biodiversity. *Nature* 405:234–42.

Chapman, F. M. 1930. Joel Asaph Allen. *Journal of Mammalogy* 11:105–17.

Cheney, D. L., and R. M. Seyfarth. 1990. *How monkeys see the world.* Chicago: University of Chicago Press.

Chesser, R. K. 1983. Genetic variability within and among populations of the black-tailed prairie dog. *Evolution* 37:320–31.

Cid, M. S., J. K. Detling, A. D. Whicker, and M. A. Brizuela. 1991. Vegetational responses of a mixed-grass prairie site following exclusion of prairie dogs and bison. *Journal of Range Management* 44:100–5.

Cincotta, R. P. 1985. Habitat and dispersal of black-tailed prairie dogs in Badlands National Park. PhD diss. Fort Collins, Colorado State University.

Cincotta, R. P., D. W. Uresk, and R. M. Hansen. 1989. Plant compositional change in a colony of black-tailed prairie dogs in South Dakota. In *Ninth Great Plains Damage Control Workshop,* 171–77. Fort Collins, CO.

Clark, T. W. 1971. Notes on white-tailed prairie dog *(Cynomys leucurus)* burrows. *Great Basin Naturalist* 31:115–25.

———. 1977. Ecology and ethology of the white-tailed prairie dog *(Cynomys leucurus). Milwaukee Public Museum Publications in Biology and Geology,* no. 3.

———. 1989. Conservation biology of the black-footed ferret. Special Scientific Report, no. 3. *Wildlife Preservation Trust International.* 1st ed.

Clark, T. W., T. M. Campbell III, D. G. Socha, and D. E. Casey. 1982. Prairie dog colony attributes and associated vertebrate species. *Great Basin Naturalist* 42:572–82.

Clark, T. W., D. Hinckley, and T. Rich. 1989. The prairie dog ecosystem: managing for biological diversity. Technical Bulletin, no. 2. *Montana BLM.*

Clutton-Brock, T., M. J. O'Riain, P.N. M/ Brotherton, D. Gaynor, R. Kansky, A. S.Griffin, and M. Manser.1999. Selfish sentinels in cooperative mammals. *Science* 284:1640–44.

Cockrum, E. L. 1960. The recent mammals of Arizona: their taxonomy and distribution. Tucson: University of Arizona Press.

Cohn, J. P. 1999. Saving the California condor—Years of effort are paying off in renewed hope for the species' survival. *Bioscience* 49:864–68.

Collier, G. D., and J. J. Spillett. 1972. Status of the Utah prairie dog *(Cynomys parvidens)*. *Utah Academy of Sciences, Arts, and Letters* 49:27–39.

Collins, A. R., J. P. Workman, and D. W. Uresk. 1984. An economic analysis of black-tailed prairie dog *(Cynomys ludovicianus)* control. *Journal of Range Management* 37:358–61.

Conner, R., A. Seidl, L. VanTassell, and N. Wilkins. 2002. *United States Grasslands and Related Resources: An economic and biological trends assessment.* College Station: Texas A&M University.

Convention on International Trade in Endangered Species of Wild Fauna and Flora. 1992. *Appendix I and II.* Châelaine-Genève, Switzerland: United Nations Environmental Program.

Cook, R. R., J. E. Carton, and P. J. Polechla, Jr. 2003. The importance of prairie dogs to nesting ferruginous hawks in grassland ecosystems. *Wildlife Society Bulletin* 31:1073–82.

Coppock, D. L., J. E. Ellis, J. K. Detling, and M. I. Dyer. 1983a. Plant-herbivore interactions in a North American mixed-grass prairie. I. Effects of black-tailed prairie dogs on intraseasonal above-ground plant biomass and nutrient dynamics and plant species diversity. *Oecologia* 56:1–9.

———. 1983b. Plant-herbivore interactions in a North American mixed-grass prairie. II. Responses of bison to modifications of vegetation by prairie dogs. *Oecologia* 56:10–15.

Costanza, R. 1996. Ecological economics: Reintegrating the study of humans and nature. *Ecological Applications* 6:978–90.

Costanza, R., R. dArge, R. deGroot, S. Farber, M. Grasso, B. Hannon, K. Limburg, S. Naeem, R. V. Oneill, J. Paruelo, R. G. Raskin, P. Sutton, and M. vandenBelt. 1997. The value of the world's ecosystem services and natural capital. *Nature* 387:253–60.

Creef, E. 1993. Greeting behavior in Gunnison's prairie dogs. MS thesis. Flagstaff: Northern Arizona University.

Cully, J. F., Jr. 1988. Gunnison's prairie dog: an important autumn raptor prey species in northern New Mexico. *National Wildlife Federation Scientific and Technical Series* 11:260–64.

———. 1991. Response of raptors to reduction of a Gunnison's prairie dog populations by plague. *American Midland Naturalist* 125:140–49.

———. 1993. Plague, prairie dogs, and black-footed ferrets. In *Proceedings of the symposium on the management of prairie dog complexes for the reintroduction of the black-footed ferret,* ed. J. L. Oldenmeyer, D. E. Biggins, and B. J. Miller, 38–49. U.S. Department of Interior Fish and Wildlife Service.

Cully, J. F., Jr, A. M. Barnes, T. J. Quan, and G. Maupin. 1997. Dynamics of plague in a Gunnison's prairie dog colony from New Mexico. *Journal of Wildlife Diseases* 33:706–19.

Cully, J. F., Jr, D. E. Biggins, and D. B. Seery. 2006. Conservation of prairie dogs in areas with plague. In *Conservation of the black-tailed prairie dog,* ed. J. L. Hoogland, 157–68. Washington, DC: Island Press.

Cully, J. F., Jr, and E. S. Williams. 2001. Interspecific comparisons of sylvatic plague in prairie dogs. *Journal of Mammalogy* 82:894–905.

Czech, B., and P. R. Krausman. 1997. Implications of an ecosystem management literature review. *Wildlife Society Bulletin* 25:667–75.

Daily, G. C., ed. 1997. *Nature's services: societal dependence on natural ecosystems.* Washington, DC: Island Press.

———. 1999. Developing a scientific basis for managing Earth's life support systems. *Conservation Ecology* 3(2):14, http://www.consecol.org/vol3/iss2/art14/

Dano, L. 1952. Cottontail rabbit *(Sylvilagus audubonii baileyi)* populations in relation to prairie dog *(Cynomys ludovicianus ludovicianus)* towns. MS thesis. Fort Collins: Colorado State University.

Davidson, A. D., R. R. Parmenter, and J. R. Gosz. 1999. Responses of small mammals and vegetation to a reintroduction of Gunnison's prairie dogs. *Journal of Mammalogy* 80:1311–24.

Davis, J. R., and T. C. Theimer. 2003. Increased lesser earless lizard *(Holbrookia maculata)* abundance on Gunnison's prairie dog colonies and short-term responses to artificial prairie dog burrows. *American Midland Naturalist* 150:282–90.

Day, T. A., and J. K. Detling. 1990a. Changes in grass leaf water relations following bison urine deposition. *American Midland Naturalist* 123:171–78.

———. 1990b. Grassland patch dynamics and herbivore grazing preference following urine deposition. *Ecology* 71:180–88.

———. 1994. Water relations of *Agropogon smithii* and *Bouteloua gracilis* and community evapotranspiration following long-term grazing by prairie dogs. *American Midland Naturalist* 132:381–92.

Derner, J. D., J. K. Detling, and M. F. Antolin. 2006. Are livestock weight gains affected by black-tailed prairie dogs? *Frontiers in Ecology and the Environment* 4:459–64.

Desmond, M. 2004. Effects of grazing practices and fossorial rodents on a winter avian community in Chihuahua, Mexico. *Biological Conservation* 116:235–42.

Desmond, M., and J. A. Savidge. 1996. Factors influencing burrowing owl nest densities and numbers in western Nebraska. *American Midland Naturalist* 136:143–48.

Desmond, M. J., J. A. Savidge, and K. M. Eskridge. 2000. Correlations between burrowing owl and black-tailed prairie dog declines: A 7-year analysis. *Journal of Wildlife Management* 64:1067–75.

Detling, J. K. 1988. Grasslands and savannas: regulation of energy flow and nutrient cycling by herbivores. In *Concepts of ecosystem ecology*, ed. J. J. Alberts and L. L. Pomeroy, 131–48. New York: Springer-Verlag.

———. 1998. Mammalian herbivores: ecosystem-level effects in two grassland national parks. *Wildlife Society Bulletin* 26:438–48.

———. 2006. Do prairie dogs compete with livestock? In *Conservation of the black-tailed prairie dog*, ed. J. L. Hoogland, 65–68. Washington, DC: Island Press.

Detling, J. K., and E. L. Painter. 1983. Defoliation responses of western wheatgrass populations with diverse histories of prairie dog grazing. *Oecologia* 57:65–71.

Detling, J. K., and A. D. Whicker. 1987. Control of ecosystem processes by prairie dogs and other grassland herbivores. *General Technical Report RM-154*. Fort Collins, CO: USDA Forest Service Rocky Mountain Range and Grassland Experimental station.

Dobson, F. S., and J. D. Kjelgaard. 1985. The influences of food resources on the life history of Columbian ground squirrels. *Canadian Journal of Zoology* 63:2105–9.

Dobson, F. S., R. K. Chesser, J. L. Hoogland, D. W. Sugg, and D. W. Foltz. 1997. Do black-tailed prairie dogs minimize inbreeding? *Evolution* 51:970–78.

———. 1998. Breeding groups and gene dynamics in a socially structured population of prairie dogs. *Journal of Mammalogy* 79:671–80.

Dunlap, T. R. 1988. *Saving America's wildlife*. Princeton, NJ: Princeton University Press.

Ecke, D. H., and C. W. Johnson. 1952. Plague in Colorado and Texas. Part I. U.S. Public Health Service. *Public Health Bulletin,* 254:83.

Egoscue, H. J., and E. S. Frank. 1984. Burrowing and denning habits of a captive colony of the Utah prairie dog. *Great Basin Naturalist* 44:495–98.

Ehrlich, P. R., and H. A. Mooney. 1983. Extinction, substitution, and ecosystem services. *Bioscience* 33:248–54.

Ehrlich, P. R., and E. O. Wilson. 1991. Biodiversity studies: science and policy. *Science* 253:758–62.

Eisner, R. J. 1991. Botanists ply trade in tropics, seeking plant-based medicines. *Scientist* 5:1.

Ellison Manning, A. E., and C. M. White. 2001. Nest site selection by mountain plovers *(Charadrius montanus)* in a shrub-steppe habitat. *Western North American Naturalist* 61:229–35.

Eske, C. R., and V. H. Haas. 1940. Plague in the western part of the United States. *Public Health Bulletin* 254:1–83.

Fagerstone, K. A., H. P. Tietjen, and O. Williams. 1981. Seasonal variation in the diet of black-tailed prairie dogs. *Journal of Mammalogy* 62:820–24.

Fahnestock, J. T., and J. K. Detling. 2002. Bison-prairie dog-plant interactions in a North American mixed-grass prairie. *Oecologia* 132:86–95.

Ferron, J. 1985. Social behavior of the golden-mantled ground squirrel *(Spermophilus lateralis)*. *Canadian Journal of Zoology* 63:2529–33.

Fitzgerald, J. P. 1970. The ecology of plague in prairie dogs and associated small mammals in South Park, Colorado. PhD diss. Fort Collins: Colorado State University.

Fitzgerald, J. P., and R. R. Lechleitner. 1974. Observations of the biology of Gunnison's prairie dog in central Colorado. *American Midland Naturalist* 92:146–63.

Flath, D. L., and R. K. Paulick. 1979. Mound characteristics of white-tailed prairie dog maternity burrow. *American Midland Naturalist* 102:395–98.

Flores, D. 1996. A long love affair with an uncommon country: Environmental history in the Great Plains. *Prairie Conservation: Preserving North America's most endangered ecosystem.* Washington, DC: Island Press.

Foltz, D. W., and J. L. Hoogland. 1981. Analysis of the mating system in the black-tailed prairie dog *(Cynomys ludovicianus)* by likelihood of paternity. *Journal of Mammalogy* 62:706–12.

Forrest, S. C., and J. C. Luchsinger. 2006. Past and current chemical control of prairie dogs. In *Conservation of the black-tailed prairie dog*, ed. J. L. Hoogland, 115–28. Washington, DC: Island Press.

Foster, K. R., P. Vecchia, and M. H. Repacholi. 2000. Risk management—Science and the precautionary principle. *Science* 288:979.

Franklin, W. L., and M. G. Garrett. 1989. Nonlethal control of prairie dog colony expansion with visual barriers. *Wildlife Society Bulletin* 17:426–30.

Frederiksen, K. 2005. A comparative analysis of alarm calls across the five species of North American prairie dogs. PhD diss. Flagstaff: Northern Arizona University.

Frederiksen, J. K., and C. N. Slobodchikoff. 2007. Referential specificity in the alarm calls of the black-tailed prairie dog. *Ethology, Ecology & Evolution* 19:87–99.

Gage, K. L., M. E. Eggleston, R. D. Gilmore, Jr, M. C. Dolan, J. A. Montenieri, D. T. Tanda, and J. Piesman. 2001. *Journal of Medical Entomology* 38:665–74.

Gage, K. L., R. S. Ostfeld, and J. G. Olson. 1995. Non-viral vector-borne zoonoses associated with mammals in the United States. *Journal of Mammalogy* 76:695–715.

Garrett, M. G. 1982. Dispersal of black-tailed prairie dogs in Wind Cave National Park, South Dakota. MS thesis. Ames: Iowa State University.

Gasper, P. W., and R. W. Watson. 2001. Plague and yersiniosis. In *Infectious diseases of wild mammals,* ed. E. S. Williams and I. K. Baker, 313–329. Ames: Iowa State University Press.

Gauthier, D. A., A. Lafon, T. P. Toombs, J. Hoth, and E. Wilken. 2003. Grasslands: Toward a North American conservation strategy. Montreal, Canada: Commission for Environmental Cooperation.

Gietzen, R. A., S. R. Jones, and R. J. McKee. 1997. Hawks, eagles, and prairie dogs: Population trends of wintering raptors in Boulder County 1983–1996. *Journal of Colorado Field Ornithologists* 31:75–89.

Gilbert-Parker, V. L. 1995. Chatter vocalizations and behavioral associations of Gunnison's prairie dogs *(Cynomys gunnisoni)* in Northern Arizona. MS thesis. Flagstaff: Northern Arizona University.

Gilligan, J. M. 2003. Precautionary principle cuts costs as well as risks. *Nature* 426:227.

Girard, J. M., D. M. Wagner, A. J. Vogler, C. Keys, C. J. Allender, L. C. Drickamer, and P. Keim. 2004. Differential plague-transmission dynamics determine *Yersinia pestis* population genetic structure on local, regional, and global scales. *Proceedings of the National Academy of Sciences USA* 101:8408–13.

Goode, G. B. 1996. Biographical sketch of Spencer Fullerton Baird. *Marine Fisheries Review* 58:40–44.

Goodrich, J. M., and S. W. Buskirk. 1998. Spacing and ecology of North American badgers *(Taxidea taxus)* in a prairie-dog *(Cynomys leucurus)* complex. *Journal of Mammalogy* 79:171–79.

Goodwin, H. T. 1995. Pliocene-Pleistocene biogeographic history of prairie dogs, genus *Cynomys* (Sciuridae). *Journal of Mammalogy* 76:100–22.

Gottelli, N. J. 1998. *A Primer of Ecology.* 2nd ed. Sunderland, MA: Sinauer and Associates.

Goulder, L. H., and D. Kennedy. 1997. Valuing ecosystem services: phiolosophical bases and empirical methods. In *Nature's services: Societal dependence on natural ecosystems,* ed. G. C. Daily, 23–47. Washington, DC: Island Press.

Grant-Hoffman, M. N., and J. K. Detling. 2006. Vegetation on Gunnison's prairie dog colonies in Southwestern Colorado. *Rangeland Ecology and Management* 59:73–79.

Green, R. A., and J. K. Detling. 2000. Defoliation-induced enhancement of total aboveground nitrogen yield of grasses. *Oikos* 91:280–84.

Hairston, N. G., F. E. Smith, and L. B. Slobodkin. 1960. Community structure, population control, and competition. *American Naturalist* 94:421–25.

Hall, E. R. 1951. *American Weasels.* Univ. Kansas Publication, Museum Natural History, vol. 4. Lawrence.

Hall, E. R., and K. Kelson. 1959. *The mammals of North America.* Vol. 1. New York: Ronald Press.

Halpin, Z. T. 1984. The role of olfactory communication in the social systems of ground-dwelling sciurids. In *The biology of ground-dwelling squirrels,* ed. J. O. Murie and G. R. Michener, 201–225. Lincoln: University of Nebraska Press.

Hamilton, W. D. 1964. The genetical evolution of social behavior. I and II. *Journal of Theoretical Biology* 7:1–52.

Hammond, A. 1998. *Which world?: Scenarios for the 21st century.* Washington, DC: Island Press.

Hanley, N., I. Moffatt, R. Faichney, and M. Wilson. 1999. Measuring sustainability: A time series of alternative indicators for Scotland. *Ecological Economics* 28:55–73.

Hansen, R. M., and I. K. Gold. 1977. Blacktail prairie dogs, desert cottontails, and cattle trophic relations on shortgrass range. *Journal of Range Management* 30:210–14.

Harlow, H. J., and S. W. Buskirk. 1996. Amino acids in plasma of fasting fat prairie dogs and lean martens. *Journal of Mammalogy* 77:407–11.

Harlow, H. J., and C. L. Frank. 2001. The role of dietary fatty acids in the evolution of spontaneous and facultative hibernation patterns in prairie dogs. *Journal of Comparative Physiology B* 171:77–84.

Harlow, H. J., and G. E. Menkens, Jr. 1986. A comparison of hibernation in the black-tailed prairie dog, white-tailed prairie dog, and Wyoming ground squirrel. *Canadian Journal of Zoology* 64:793–96.

Haynie, M. L., R. A. Van Den Bussche, J. L. Hoogland, and D. A. Gilbert. 2003. Parentage, multiple paternity, and breeding success in Gunnison's and Utah prairie dogs. *Journal of Mammalogy* 84:1244–53.

Heffner, R. S., H. E. Heffner, C. Contos, and D. Kearns. 1994. Hearing in prairie dogs: transition between surface and subterranean rodents. *Hearing Research* 73:185–89.

Heimlich, R. E., and W. D. Anderson. 2001. Development at the urban fringe and beyond: impacts on agricultural and rural land. In *Agricultural Economic Report 803,* Economic Research Service, U.S. Department of Agriculture. Washington, DC: Government Printing Office.

Henwood, W. 1998. An overview of protected areas in the temperate grasslands biome. *Parks* 8:3–8.

Herron, M. D., T. A. Castoe, and C. L. Parkinson. 2004. Sciurid phylogeny and the paraphyly of Holarctic ground squirrels *(Spermophilus). Molecular Phylogenetics and Evolution* 31:1015–30.

Hillman, C. N. 1968. Field observation of Black-footed ferrets in South Dakota. *Transactions North American Wildlife and Natural Resources Conference Proceedings* 33:433–43.

Hockett, C. F. 1960. Logical considerations in the study of animal language. In *Animal sounds and communication*, ed. W. E. Lanyon and W. N. Tavolga. *Publication No. 7*, 392–430. Washington, DC: American Institute of Biological Sciences.

Holdenried, R., and S. F. Quan. 1956. Susceptibility of New Mexico rodents to experimental plague. *Public Health Report* 71:979–94.

Holland, E. A., and J. K. Detling. 1990. Plant response to herbivory and below ground nitrogen cycling. *Ecology* 71:1040–49.

Holycross, A. T., and J. D. Fawcett. 2002. Observations on neonatal aggregations and associated behaviors in the prairie rattlesnake, *Crotalus viridis viridis*. *American Midland Naturalist* 148:181–84.

Hoogland, J. L. 1979. The effect of colony size on individual alertness of prairie dogs: Sciuridae *Cynomys* spp. *Animal Behaviour* 27:394–407.

———. 1981. The evolution of coloniality in white-tailed and black-tailed prairie dogs (Sciuridae: *Cynomys leucurus* and *C. ludovicianus*. *Ecology* 62:252–72.

———. 1982. Prairie dogs avoid extreme inbreeding. *Science* 215:1639–41.

———. 1983. Nepotism and alarm calling in the black-tailed prairie dog *(Cynomys ludovicianus)*. *Animal Behaviour* 31:472–79.

———. 1985. Infanticide in prairie dogs: Lactating females kill offspring of close kin. *Science* 230:1037–40.

———. 1992. Levels of inbreeding among prairie dogs. *American Naturalist* 139:591–602.

———. 1995. *The black-tailed prairie dog: social life of a burrowing mammal*. Chicago: University of Chicago Press.

———. 1996. Why do Gunnison's prairie dogs give alarm calls? *Animal Behaviour* 51:871–80.

———. 1997. Duration of gestation and lactation for Gunnison's prairie dogs. *Journal of Mammalogy* 78:173–80.

———. 1998a. Estrus and copulation for Gunnison's prairie dogs. *Journal of Mammalogy* 79:887–97.

———. 1998b. Why do Gunnison's prairie dog females copulate with more than one male? *Animal Behaviour* 55:351–59.

———. 1999. Philopatry, dispersal, and social organization of Gunnison's prairie dogs. *Journal of Mammalogy* 80:243–51.

———. 2001. Black-tailed, Gunnison's, and Utah prairie dogs reproduce slowly. *Journal of Mammalogy* 82:917–27.

———. 2006. Saving prairie dogs: Can we? Should we? In *Conservation of the black-tailed prairie dog*, ed. J. L. Hoogland, 261–266. Washington, DC: Island Press.

Hoogland, J. L., D. K. Angell, J. G. Daley, and M. C. Radcliffe. 1987. Demography and population dynamics of prairie dogs. In Eighth Great Plains wildlife damage control workshop. Rapid City, South Dakota.

Hoogland, J. L., S. Davis, S. Benson-Amram, D. Labruna, B. Goossens, and M. A. Hoogland. 2004. Pyraperm kills fleas and halts plague among Utah prairie dogs. *Southwestern Naturalist* 49:376–83.

Hoogland, J. L., R. H. Tamarin, and C. K. Levy. 1989. Communal nursing in prairie dogs. *Behavioral Ecology and Sociobiology* 24:91–95.

Hrdy, S. B., and G. Hausfater. 1984. Comparative and evolutionary perspectives on infanticide: Introduction and overview. In *Infanticide: Comparative and evolutionary perspectives*, ed. G. Hausfater and S. B. Hrdy, xiii–xxxv. New York: Aldine.

Humphrey, R. R. 1958. *The desert grassland.* Tuscon: University of Arizona Press.

Hunter, M. D., and P. W. Price. 1992. Playing chutes and ladders: heterogeneity and the relative roles of bottom-up and top-down forces in natural communities. *Ecology* 73:724–32.

Hygnstrom, S. E., and K. C. VerCauteren. 2000. Cost-effectiveness of five burrow fumigants for managing black-tailed prairie dogs. *International Biodeterioration and Biodegradation* 45:159–68.

Ingham, R. E., and J. K. Detling. 1984. Plant herbivore interactions in a North American mixed grass prairie III. Soil nematode populations and root biomass on *Cynomys ludovicianus* colonies and adjacent uncolonized areas. *Oecologia (Berlin)* 63:307–13.

Jackson, J. A. 2000. Baird, Spencer Fullerton. *American National Biography Online,* www.anb.org.

Jacobs, G. H. 1981. *Comparative color vision.* New York: Academic Press.

Jacobs, G. H., and K. A. Pulliam. 1973. Vision in the prairie dog: Spectral sensitivity and color vision. *Journal of Comparative and Physiological Psychology* 84:240–45.

Jaramillo, V. J., and J. K. Detling. 1988. Grazing history, defoliation, and competition: Effects on shortgrass production and nitrogen accumulation. *Ecology* 69:1599–1608.

Johnson, W. C., and S. K. Collinge. 2004. Landscape effects on prairie dog colonies. *Biological Conservation* 115:487–97.

Jones, C. C., J. H. Lawton, and M. Shachak. 1994. Organisms as ecosystem engineers. *Oikos* 69:373–86.

Jones, S. R. 1989. Populations and prey selection of wintering raptors in Boulder County, Colorado. *Proceedings of the North American Prairie Conference* 11:255–58.

Jones, T. R., and R. K. Plakke. 1981. The histology and histochemistry of the perianal scent gland of the reproductively quiescent black-tailed prairie dog *(Cynomys ludovicianus). Journal of Mammalogy* 62:362–68.

Journal of the American Veterinary Medical Association (JAVMA). 2002. News Archive, October 1, 2002. "Tularemia outbreak identified in pet prairie dogs." http://www.avma.org/onlnews/javma/oct02/021001g.asp

Kahn, P. H., Jr. 2001. *The Human Relationship with Nature.* Cambridge, MA: MIT Press.

Kahn, P. H., Jr, and S. R. Kellert. 2002. *Children and Nature: Psychological, Sociocultural, and Evolutionary Investigations.* Cambridge, MA: MIT Press.

Kaplan, R., and S. Kaplan. 1989. *The experience of nature: a psychological perspective.* Cambridge: Cambridge University Press.

Kaufmann, A. F., J. M. Mann, T. M. Gardiner, F. Heaton, J. D. Poland, A. M. Barnes, and G. O. Maupin. 1981. Public health implications of plague in domestic cats. *Journal of the American Veterinary Medical Association* 179:875–78.

Keeling, M. J., and C. A. Gilligan. 2000. Metapopulation dynamics of plague. *Nature* 6806:903–6.

Kellert, S. R., and E. O. Wilson. 1993. *The Biophilia Hypothesis.* Washington, DC: Island Press.

Kelso, L. H. 1939. Food habits of prairie dogs. *U.S.D.A. Circular* 529:1–15.

Kildaw, S. B. 1995. The effect of group size manipulations on the foraging behavior of black-tailed prairie dogs. *Behavioral Ecology* 6:353–58.

Kimberling, C. 2006. Constantine Samuel Rafinesque (1783–1840), naturalist. http://faculty.evansville.edu/ck6/bstud/rafin.html.

King, J. A. 1955. Social behavior, social organization, and population dynamics in a black-tailed prairiedog town in the Black Hills of South Dakota. *Contributions from the Laboratory of Vertebrate Biology, University of Michigan* 67:1–123.

King, J. A. 1984. Historical ventilations on a prairie dog town. In *The biology of ground-dwelling squirrels*, ed. J. O. Murie and G. R. Michener, 447–56. Lincoln: University of Nebraska Press.

Kiriazis, J. 1991. Communication and sociality in Gunnison's prairie dogs. PhD diss. Flagstaff: Northern Arizona University.

Kiriazis, J., and C. N. Slobodchikoff. 2006. Perceptual specificity in the alarm calls of Gunnison's prairie dogs. *Behavioural Processes* 73:29–35.

Knowles, C. 1982. Habitat affinity, populations, and control of black-tailed prairie dogs on the Charles M. Russell National Wildlife Refuge. PhD diss. Missoula: University of Montana.

———. 1985. Observations on Prairie Dog dispersal in Montana. *Prairie Naturalist* 17:33–40.

———. 1986. Some relationships of black-tailed prairie dogs to livestock grazing. *Great Basin Naturalist* 46:198–203.

———. 1987. Reproductive ecology of black-tailed prairie dogs in Montana. *Great Basin Naturalist* 47:202–6.

———. 1988. An evaluation of shooting and habitat alteration for control of black-tailed prairie dogs. *Proceedings of the Great Plains Wildlife Damage Control Conference* 8:53–56.

Knowles, C. J., and P. R. Knowles 1994. A review of black-tailed prairie dog literature in relation to rangelands administered by the Custer National Forest. Washington, DC: *U.S. Forest Service.*

Knowles, C. J., J. Proctor, and S. C. Forrest. 2002. Black-tailed prairie dog abundance and distribution on the Northern Great Plains based on historic and contemporary information. *Great Plains Research* 12:219–54.

Knowles, C. J., C. J. Stoner, and S. P. Gieb. 1982. Selective use of black-tailed prairie dog towns by mountain plovers. *Condor* 84:71–74.

Koford, C. B. 1958. Prairie dogs, whitefaces, and blue grama. *Wildlife Monographs* 3:1–78.

Kolbe, J. J., B. E. Smith, and D. M. Browning. 2002. Burrow use by tiger salamanders *(Ambystoma tigrinum)* at a black-tailed prairie dog *(Cynomys ludovicianus)* town in southwestern South Dakota. *Herpetological Review* 33:35–99.

Kotliar, N. B. 2000. Application of the keystone species concept to prairie dogs: How well does it work? *Conservation Biology* 14:1715–21.

Kotliar, N. B., B. W. Baker, A. D. Whicker, and G. Plumb. 1999. A critical review of assumptions about the prairie dog as a keystone species. *Environmental Management* 24:177–92.

Kotliar, N. B., B. J. Miller, R. P. Reading, and T. W. Clark. 2006. The prairie dog as keystone species. In *Conservation of the black-tailed prairie dog*, ed. J. L. Hoogland, 53–64. Washington, DC: Island Press.

Kretzer, J. E., and J. F. Cully. 2001. Prairie dog effects on harvester ant species diversity and density. *Journal of Range Management* 54:11–14.

Kretzer, J. E., and J. F. Cully, Jr. 2001. Effects of black-tailed prairie dogs on reptiles and amphibians in Kansas shortgrass prairie. *Southwestern Naturalist* 46:171–177.

Krueger, K. 1986. Feeding relationships among bison, pronghorn, and prairie dogs: An experimental analysis. *Ecology* 67:760–70.

Kuchler, A. W. 1985. Potential national vegetation in national atlas of the United States of America. U.S. Department of Interior, United States Geological Survey, Reston.

Kuffner, C. 2002. Off the Kuff. August 5, 2002, It's a Prairie Dogs Life. http://www.offthekuff.com/mt/archives2/2002/08

Lamb, B. L., R. P. Reading, and W. F. Andelt. 2006. Attitudes and perceptions about prairie dogs. In *Conservation of the black-tailed prairie dog*, ed. J. L. Hoogland, 108–114. Washington, DC: Island Press.

Lawn, P. A. 2003. A theoretical foundation to support the Index of Sustainable Economic Welfare (ISEW), Genuine Progress Indicator (GPI), and other related indexes. *Ecological Economics* 44:105–118.

Lawton, J. H. 1994. Parasitoids as model communities in ecological theory. In *Parasitoid Community Ecology*, ed. B. A. Hawkins and W. Sheehan, 492–506. Oxford University Press.

Lechleitner, R. R., L. Kartman, M. I. Goldenberg, and B. W. Hudson. 1968. An epizootic of plague in Gunnison's prairie dogs *(Cynomys gunnisoni)* in south-central Colorado. *Ecology* 49:734–43.

Lehmer, E. M., and D. E. Biggins. 2005. Variation in torpor patterns of free-ranging black-tailed and Utah prairie dogs across gradients of elevation. *Journal of Mammalogy* 86:15–21.

Lehmer, E. M., and B. Van Horne. 2001. Seasonal changes in lipids, diet, and body composition of free-ranging black-tailed prairie dogs *(Cynomys ludovicianus)*. *Canadian Journal of Zoology* 79:955–65.

Lehmer, E. M., B. Van Horne, B. Kulbartz, and G. L. Florant. 2001. Facultative torpor in free-ranging black-tailed prairie dogs *(Cynomys ludovicianus)*. *Journal of Mammalogy* 82:551–57.

Lewis, J. K. 1982. Use of ecosystem classification in range resource management. In *Grassland Ecology and Classification Synposium proceedings*, ed. A. C. Nicholson, A. McClean, and T. E. Baker. Ministry of Forests, Province of British Columbia.

Lidicker, W. C., and N. C. Stenseth. 1992. *Animal dispersal: small mammals as a model.* New York, London: Chapman and Hall.

Lima, S. L. 1995. Back to the basics of anti-predatory vigilance: the group size effect. *Animal Behaviour* 49:11–20.

Linder, R. L., R. B. Dahlgren, and C. N. Hillman. 1972. Black-footed ferret-prairie dog interrelationships. In *Symposium on Rare and Endangered Wildlife of the Southwestern United States; September 22–23, 1972,* 22–37. Albuquerque: New Mexico Department of Game and Fish.

List, R., and D. W. MacDonald. 2003. Home range and habitat use of the kit fox *(Vulpes macrotis)* in a prairie dog *(Cynomys ludovicianus)* complex. *Journal of Zoology (London)* 259:1–5.

Lomolino, M. V., and G. A. Smith. 2003. Prairie dog towns as islands: applications of island biogeography and landscape ecology for conserving nonvolant terrestrial vertebrates. *Global Ecology and Biogeography* 12:275–86.

———. 2004. Terrestrial vertebrate communities at black-tailed prairie dog *(Cynomys ludovicianus)* towns. *Biological Conservation* 115:89–100.

Long, D., K. Bly-Honness, J. C. Truett, and D. B. Seery. 2006. Establishment of new prairie dog colonies by translocation. In *Conservation of the black-tailed prairie dog,* ed. J. L. Hoogland, 188–209. Washington, DC: Island Press.

Long, S. T. 1998. Alarm calls of Gunnison's prairie dogs *(Cynomys gunnisoni)* reflect levels of threat. MS thesis. Flagstaff: Northern Arizona University.

Longhurst, W. 1944. Observations on the ecology of the Gunnison's prairie dog in Colorado. *Journal of Mammalogy* 25:24–36.

Loreau, M. 2000. Biodiversity and ecosystem function: recent theoretical advances. *Oikos* 91:3–17.

Loughry, W. J. 1987a. The dynamics of snake harassment by black-tailed prairie dogs. *Behaviour* 103:27–48.

———. 1987b. Differences in natural and experimental encounters of black-tailed prairie dogs with snakes. *Animal Behaviour* 35:1568–70.

———. 1988. Population differences in how black-tailed prairie dogs deal with snakes. *Behavioral Ecology and Sociobiology* 22:61–67.

———. 1989. Discrimination of snakes by two populations of black-tailed prairie dogs. *Journal of Mammalogy* 70:627–30.

Loughry, W. J., and A. Lazari. 1994. The ontogeny of individuality in the black-tailed prairie dogs, *Cynomys ludovicianus. Canadian Journal of Zoology* 72:1280–86.

Lowe, C. 1985. *Arizona's natural environment.* Tucson: University of Arizona Press.

Luce, R. J., R. Manes, and B. Van Pelt. 2006. A multi-state plan to conserve prairie dogs. In *Conservation of the black-tailed prairie dog*, ed. J. L. Hoogland, 210–17. Washington, DC: Island Press.

Ludwig, D., M. Mangel, and B. Haddad. 2001. Ecology, conservation, and public policy. *Annual Review of Ecology and Systematics* 32:481–517.

MacArthur, R. H., and E. O. Wilson. 1967. The theory of island biogeography. *Monographs in Population Biology,* no. 1. New Jersey: Princeton University Press.

Macedonia, J. M., and C. S. Evans. 1993. Variation among mammalian alarm call systems and the problem of meaning in animal signals. *Ethology* 93:177–97.

Manes, R. 2006. Does the prairie dog merit protection via the Endangered Species Act? In *Conservation of the black-tailed prairie dog*, ed. J. L. Hoogland, 169–83. Washington, DC: Island Press.

Manzano-Fischer, P. 1996. Avian communities associated with prairie dog towns in northwestern Mexico. MS thesis, Oxford University.

Manzano-Fischer, P., R. List, and G. Ceballos. 1999. Grassland birds in prairie dog towns in northwestern Chihuahua, Mexico. *Studies in Avian Biology* 19:263–71.

Marsh, R. E. 1984. Ground squirrels, prairie dogs and marmots as pests on rangeland. In *Proceedings of the conference for organization and practice of vertebrate pest control*, 195–208. Fernherst, England: ICI Plant protection division.

Mayr, E. 1969. The biological meaning of species. *Biological Journal of the Linnean Society* 1:311–20.

McCracken, R. J., J. S. Lee, R. W. Arnold, and D. E. McCormack. 1985. *An appraisal of soil resources in the USA.* Madison, WI: American Society of Agronomy.

McCullough, D. A., and R. K. Chesser. 1987. Genetic variation among populations of the Mexican prairie dog. *Journal of Mammalogy* 68:555–60.

McCullough, D. A., R. K. Chesser, and R. D. Owens. 1987. Immunological systematics of prairie dogs. *Journal of Mammalogy* 68:561–68.

Mellink, E., and H. Madrigal. 1993. Ecology of Mexican prairie dogs *(Cynomys mexicanus)* in El Manantial, Northeastern Mexico. *Journal of Mammalogy* 74:631–35.

Mencher, J. S., S. R. Smith, T. D. Powell, D. T. Stinchcomb, J. E. Osorio, and T. E. Rocke. 2004. Protection of black-tailed prairie dogs *(Cynomys ludovicianus)* against plague after voluntary consumption of baits containing recombinant raccoon poxvirus vaccine. *Infection and Immunity* 72:5502–5.

Menkens, G. E., Jr., 1987. Temporal and spatial variation in white-tailed prairie dog *(Cynomys leucurus)* populations and life histories in Wyoming. PhD diss. Laramie: University of Wyoming.

Menkens, G. E., Jr., and S. H. Anderson. 1989. Temporal-Spatial variation in white-tailed prairie dog demography and life histories in Wyoming. *Canadian Journal of Zoology* 67:343–49.

———. 1991. Population dynamics of white-tailed prairie dogs during an epizootic of sylvatic plague. *Journal of Mammalogy* 72:328–31.

Merriam, C. H. 1902. The prairie dog of the great plains. In *Yearbook of the United States Department of Agriculture (1901)*, 257–70. Washington, DC.

Merrill, E. D. 1949. *Index Rafinesquianus.* Cambridge, MA: The Arnold Arboretum, Harvard University.

Milchunas, D. G. 2006. Responses of plant communities to grazing in the southwestern United States. *USDA General Technical Report RMRS-GTR-169.* Fort Collins, CO.

Milchunas, D. G., O. E. Sala, and W. K. Lauenroth. 1988. A generalized model of the effects of grazing by large herbivores on grassland community structure. *American Naturalist* 132:87–106.

Miles, V. I., M. J. Wilcomb, and J. V. Irons. 1952. Plague in Colorado and Texas. Part II. Rodent plague in the Texas south plains 1947–1949 with ecological considerations. *Public Health Monographs* 6:41–53.

Miller, B., D. Biggins, and R. Crete. 1993. Workshop summary. In *Proceedings of the symposium on the management of prairie dog complexes for the reintroduction of the black-footed ferret,* ed. J. L. Oldenmeyer et al. U.S. Department of the Interior, Fish and Wildlife Service.

Miller, B., and G. Ceballos. 1994. Managing conflict and biotic diversity in the prairie dog ecosystem. *Endangered Species Update* 11:1–4.

Miller, B., G. Ceballos, and R. Reading. 1994. The prairie dog and biotic diversity. *Conservation Biology* 8:677–81.

Miller, B., and R. Reading. 2006. A proposal for more effective conservation of prairie dogs. In *Conservation of the black-tailed prairie dog,* ed. J. L. Hoogland, 248–60. Washington, DC: Island Press.

Miller, B., R. Reading, and S. Forrest. 1996. *Prairie night: Black-footed ferrets and the recovery of endangered species.* Washington, DC: Smithsonian Institution Press.

Miller, S. D., R. Reading, B. Haskins, and D. Stern. 2005. Overestimation bias in estimates of black-tailed prairie dog abundance in Colorado. *Wildlife Society Bulletin* 33:1444–51.

Miller, B., R. Reading, J. Hoogland, T. Clark, G. Ceballos, R. List, S. Forrest, L. Hanebury, P. Manzano, J. Pacheco, and D. Uresk. 2000. The role of prairie dogs as keystone species: response to Stapp. *Conservation Biology* 14:318–21.

Miller, B., C. Wemmer, D. Biggins, and R. Reading. 1990. A proposal to conserve black-footed ferrets and the prarie dog ecosystem. *Environmental Management* 14:763–69.

Mills, E. S., and P. E. Graves. 1986. Economics and environmental quality: The basics. In *The economics of environmental quality*, ed. E. S. Mills and P. E. Graves, 28–49. New York: W.W. Norton.

Mills, L. S., M. E. Soule, and D. Doak. 1993. The keystone species concept in ecology and conservation. *BioScience* 43:219–24.

Mooney, H. A., and P. R. Ehrlich. 1997. Ecosystem services: a fragmentary history. In *Nature's services: Societal dependence on natural ecosystems,* ed. G. C. Daily, 11–19. Washington, DC: Island Press.

Moore, E. 2002. "The Prairie Dog Lady." Texas Magazine. *Sunday Houston Chronicle,* December 1, 8.

Morowitz, H. J. 1991. Balancing species preservation and economic considerations. *Science* 253:752–754.

Morton, M. L., and P. W. Sherman. 1978. Effects of a spring snow storm on behavior, reproduction, and survival of Belding's ground squirrels. *Canadian Journal of Zoology* 56:2578–90.

Motiff, J. P. 1980. Frequency of alarm barks in two black-tailed prairie dog towns. *Psychological Reports* 46:1164–66.

Mulhern, D. W., and C. J. Knowles. 1995. Black-tailed prairie dog status and future conservation planning. *General Technical Report RM-GTR-298,* United States Department of Agriculture Forest Service. Fort Collins, CO: Rocky Mountain Forest Range Experimental Station.

Munn, L. C. 1993. Effects of prairie dogs on physical and chemical properties of soils. In *Proceedings of the symposium on the management of prairie dog complexes for the reintroduction of the black-footed ferret,* ed. J. L. Oldenmeyer, et al. 11–17. U.S. Department of Interior, U.S. Fish and Wildlife Service.

Murie, J. O. 1985. A comparison of life history traits in two populations of *Spermophilus columbianus* in Alberta Canada. *Acta Zoologica Fennica* 173:43–45.

Myers, N., and J. Kent. 1998. Perverse subsidies: taxes undercutting our economies and environments alike. In International Institute for Sustainable Development, ed. N. Myers. Oxford, UK. On-line resource: http://www.brocku.ca/envi/db/envi1p90/readings/Perverse%20Subsidies%20Executive%20Summary.pdf#search='perverse%20subsidies%20definition

Nadler, C. F., R. S. Hoffman, and J. J. Pizzimenti. 1971. Chromosomes and serum proteins of prairie dogs and a model of *Cynomys* evolution. *Journal of Mammalogy* 52:545–55.

Nott, M. P., E. Rogers, and S. Pimm. 1995. Extinction rates—modern extinction rates in the kilo-death range. *Current Biology* 5:14–17.

Odum, H. T. 1996. *Environmental accounting: EMERGY and decisionmaking.* New York: John Wiley.

Oldenmeyer, J. L., D. E. Biggins, and B. J. Miller, ed. 1993. *Proceedings of the symposium on the management of prairie dog complexes for the reintroduction of the black-footed ferret.* U.S. Department of the Interior, Fish and Wildlife Service.

Olson, S. L. 1984. Density and distribution, nest site selection and activity of the mountain plover on the Charles M Russell National Wildlife Refuge. MS thesis. Missoula: University of Montana.

Olson, S. L., and W. D. Edge. 1985. Nest site selection by mountain plovers in northcentral Montana. *Journal of Range Managment.* 38:280–82.

O'Meilia, M. E., F. L. Knopf, and J. C. Lewis. 1982. Some consequences of competition between prairie dogs and beef cattle. *Journal of Range Management* 35:580–85.

Osgood, W. H. 1943. Clinton Hart Merriam, 1855–1942. *Journal of Mammalogy* 24:421–36.

Owings, D. H., and D. F. Hennessy. 1984. The importance of variation in sciurid visual and vocal communication. In *The biology of ground-dwelling squirrels,* ed. J. O. Murie and G. R. Michener, 169–200. Lincoln: University of Nebraska Press.

Owings, D. H., and W. J. Loughry. 1985. Variation in snake-elicited jump-yipping by black-tailed prairie dogs: Ontogeny and snake-specificity. *Zeitschrift fur Tierpsychologie* 70:177–200.

Owings, D. H., and E. S. Morton. 1998. *Animal vocal communication: A new approach.* New York: Cambridge University Press.

Owings, D. H., and S. C. Owings. 1979. Snake-directed behavior by black-tailed prairie dogs *(Cynomys ludovicianus). Zeitschrift fur Tierpsychologie* 49:35–54.

Paine, R. T. 1969. A note on trophic complexity and community stability. *American Naturalist* 103:91–93.

Painter, E. L., J. K. Detling, and D. A. Steingraeber. 1993. Plant morphology and grazing history: relationships between native grasses and herbivores. *Vegetatio* 106:37–62.

Pauli, J. N. 2005. Ecological studies of the black-tailed prairie dog *(Cynomys ludovicianus):* Implications for biology and conservation. MS thesis. Laramie: University of Wyoming.

Pauli, J. N., and S. W. Buskirk. 2007. Recreational shooting of prairie dogs: A portal for lead entering wildlife food chains. *Journal of Wildlife Management* 71:103–8.

Peck, R. M. 2000. Ord, George. American National Biography Online, www.anb.org.

Perla, B., and C. N. Slobodchikoff. 2002. Habitat structure and alarm call dialects in the Gunnison's prairie dog *(Cynomys gunnisoni)*. *Behavioral Ecology* 13:844–50.

Pizzimenti, J. J., and L. R. McClenaghan. 1974. Reproduction, growth and development, and behavior in the Mexican prairie dog. *American Midland Naturalist* 92:130–45.

Placer, J., and C. N. Slobodchikoff. 2000. A fuzzy-neural system for identification of species-specific alarm calls of Gunnison's prairie dogs. *Behavioural Processes* 52:1–9.

———. 2001. Developing new metrics for the investigation of animal vocalizations. *Intelligent Automation and Soft Computing* 7:1–11.

———. 2004. A method for identifying sounds used in alarm call classification. *Behavioural Processes* 67:87–98.

Poland, J. D., and A. M. Barnes. 1979. Plague. In *CRC handbook series in zoonoses, Section A. Bacterial rickettsial and myotic diseases* ed. J. H. Steele, 1:515–56.

Polderboer, E. B., L. W. Kuhn, and G. O. Hendrickson. 1941. Winter and spring habitats of weasels in central Iowa. *Journal of Wildlife Management* 5:115–19.

Polley, H. W., and J. K. Detling. 1989. Defoliation, Nitrogen and competition: effects on plant growth and nitrogen nutrition. *Ecology* 70:721–27.

Pollitzer, R. 1954. Plague. *World Health Organization Health Series 22*. Geneva, Switzerland: World Health Organization.

Portney, P. R. 1994. The contingent valuation debate: why economists should care. *Journal of Economic Perspectives* 8:2–17.

Powell, K. L. 1992. Prairie dog distribution, habitat characteristics, and population monitoring in Kansas: implications for black-footed ferret recovery. MS thesis. Manhattan: Kansas State University.

Powell, K. L., R. J. Robel, K. E. Kemp, and M. D. Nellis. 1994. Aboveground counts of black-tailed prairie dogs: Temporal nature and relationship to burrow-entrance density. *Journal of Wildlife Management* 58:361–66.

Power, M. E., D. Tilman, J. A. Estes, B. A. Menge, W. J. Bond, L. S. Mills, G. Dailt, J. C. Castilla, J. Lubchenco, and R. T. Paine. 1996. Challenges in the quest for keystones. *Bioscience* 466:9–20.

Price, E. R. 2002. Observation of rock wrens using white-tailed prairie dog burrows. *Prairie Naturalist* 34:149–50.

Price, M. V., and O. J. Reichman. 1987. Distribution of seeds in Sonoran desert soils: Implications for heteromyid rodent foraging. *Ecology* 68:1797–1811.

Proctor, J., B. Haskins, and S. C. Forrest. 2006. Focal areas for conservation of prairie dogs and the grassland ecosystem. In *Conservation of the black-tailed prairie dog*, ed. J. L. Hoogland, 232–47. Washington, DC: Island Press.

Prugh, T., and E. Assadourian. 2003. What is sustainability anyway? *WorldWatch* 16:10–12.

Rayor, L. S. 1985. Effects of habitat quality on growth, age of first reproduction and dispersal in Gunnison's prairie dogs. *Canadian Journal of Zoology.* 63:2835–40.

———. 1988. Social organization and space-use in Gunnison's prairie dog. *Behavioral Ecology and Sociobiology* 22:69–78.

Rayor, L. S., A. K. Brody, and C. Gilbert. 1987. Hibernation in the Gunnison's prairie dog. *Journal of Mammalogy* 68:147–50.

Reading, R. P., B. J. Miller, and S. R. Kellert. 1999. Values and attitudes toward prairie dogs. *Anthrozoos* 12:43–50.

Reeve, A. F., and T. C. Vosburgh. 2006. Recreational shooting of prairie dogs. In *Conservation of the black-tailed prairie dog,* ed. J. L. Hoogland, 139–56. Washington, DC: Island Press.

Reichman, O. J. 1984. Spatial and temporal variation of seed distributions in Sonoran Desert soils. *Journal of Biogeography* 11:1–11.

Restani, M., L. R. Rau, and D. L. Flath. 2001. Nesting ecology of burrowing owls occupying black-tailed prairie dog towns in southeastern Montana. *Journal of Raptor Research* 35:296–303.

Reveal, J. L. 2003. Constantine Samuel Rafinesque-Schmaltz (1783–1840). www. lewis-clark.org.

Rice, R. E. 1990. Old growth logging myths: the ecological impact of the U.S. Forest Service's management policies. *Ecologist* 20:141.

Ricketts, T. H., E. Dinerstein, D. M. Olson, C. J. Loucks, W. Eichbaum, D. DellaSala, K. Kavanagh, P. Hedao, P. T. Hurley, K. M. Carney, R. Abell, and S. Walters. 1999. *Terrestrial ecoregions of North America: A conservation assessment.* Washington, DC: Island Press.

Riebsane, W. E., H. Gosnell, and D. M. Theobold. 1996. Lands use and landscape change in the Colorado Mountains. I: Theory, scale, and pattern. *Mountain Research and Development* 16:395.

Roach, J. L., P. Stapp, B. Van Horne, and M. Antolin. 2001. Genetic structure of a metapopulation of black-tailed prairie dogs. *Journal of Mammalogy* 82:946–59.

Robinson, A. T. 1989. Dispersal of the Gunnison's prairie dog, *Cynomys gunnisoni.* MS thesis. Flagstaff: Northern Arizona University.

Roemer, D. M., and S. C. Forrest. 1996. Prairie dog poisoning in Northern Great Plains: an analysis of programs and policies. *Environmental Management* 20:349–59.

Rosmarino, N. J. 2006. Box 12.1. Why the prairie dog merits ESA protection. In *Conservation of the black-tailed prairie dog,* ed. J. L. Hoogland, 178–80. Washington, DC: Island Press.

Sala, O. E., and J. M. Paruelo. 1997. Ecosystem services in grasslands. In *Nature's services: societal dependence on natural ecosystems,* ed. G. C. Daily, 237–52. Washington, DC: Island Press.

Sampson, F. B., and F. L. Knopf. 1994. Prairie Conservation in North America. *BioScience* 44:418–21.

Scheffer, T. H. 1937. Study of a small prairie dog town. *Transactions of the Kansas Academy of Sciences* 40:391–94.

Schloemer, R. D. 1991. Prairie dog effects on vegetation and soils derived from shale in Shirley Basin, Wyoming. PhD diss. Laramie: University of Wyoming.

Schroeder, M. 1988. Endangered species considerations in prairie dog management. *U.S. Forest Service General Technical Report RM 154,* 123–24.

Schwartz, C. C., and J. E. Ellis. 1981. Feeding ecology and niche separation in some native and domestic ungulates on the short-grass prairie. *Journal of Applied Ecology* 18:343–53.

Seery, D. B., D. E. Biggins, J. A. Montenieri, R. E. Enscore, D. T. Tanda, and K. L. Gage. 2003. Treatment of black-tailed prairie dog burrows with deltamethrin to control fleas (Insecta: Syphonaptera) and plague. *Journal of Medical Entomology* 40:718–31.

Seglund, A. E., A. E. Ernst, M. Grenier, B. Luce, A. Puchniak, and P. Schnurr. 2006a. *White-tailed prairie dog conservation assessment.* Western Association of Fish and Wildlife Agencies.

Seglund, A. E., A. E. Ernst, and D. M. O'Neill. 2006b. *Gunnison's prairie dog conservation assessment.* Western Association of Fish and Wildlife Agencies.

Seton, E. T. 1929. *Lives of game animals.* Garden City, NY: Doubleday Dovan Inc.

Severson, K. E., and G. E. Plumb. 1998. Comparison of methods to estimate population densities of black-tailed prairie dogs. *Wildlife Society Bulletin* 26:859–66.

Shalaway, S., and C. N. Slobodchikoff. 1988. Seasonal changes in the diet of Gunnison's prairie dogs. *Journal of Mammalogy* 69:835–41.

Sharps, J. 1988. Politics, prairie dogs, and the sportsman. In *Eighth Great Plains Wildlife Damage Control Workshop Proceedings. U.S. Forest Service Gen. Tech. Rep. RM-154,* ed D. W. Uresk and G. Schenbeck, 117–118. Fort Collins, CO: USFS Rocky Mountain Forest and Range Experimental Station.

Sharps, J. C., and D. W. Uresk. 1990. Ecological review of black-tailed prairie dogs and associated species in western South Dakota. *Great Basin Naturalist* 50:339–345.

Sheets, R. G., R. L. Linder, and R. B. Dahlgren. 1971. Burrow systems of prairie dogs in South Dakota. *Journal of Mammalogy* 52:451–53.

Sheffield, S. R., and M. Howery. 2001. Current status, distribution, and conservation of the burrowing owl in Oklahoma. *Journal of Raptor Research* 35:351–56.

Shier, D. M. 2006. Box 13.1. Translocations are more successful when prairie dogs are moved as families. In *Conservation of the black-tailed prairie dog,* ed. J. L. Hoogland, 189–190. Washington, DC: Island Press.

Shipley, B., and R. P. Reading. 2006. A comparison of herpetofauna and small mammal diversity on black-tailed prairie dog (*Cynomys ludovicianus*) colonies and non-colonized grassland in Colorado. *Journal of Arid Environments* 66:27–41.

Shogren, J. F. 2000. Economics and the Endangered Species Act, USA. Online report: http://www.umich.edu/~esupdate/library/97.01–02/shogren.html. Laramie: University of Wyoming.

Sidle, J. G., M. Ball, T. Byer, J. J. Chynoweth, G. Foli, R. Hodorff, G. Moravek, R. Peterson, and D. J. Svingen. 2001. Occurrence of burrowing owls in black-tailed prairie dog colonies on Great Plains National Grasslands. *Journal of Raptor Research* 35:316–21.

Slobodchikoff, C. N., ed. 1976. *Concepts of species.* Stroudsburg, PA: Dowden, Hutchinson & Ross.

———. 1984. Resources and the evolution of social behavior. In *A new ecology: Novel approaches to interactive systems,* ed. P. W. Price, C. N. Slobodchikoff, and W. Gaud, 227–51. New York: John Wiley and Sons.

———. 1987. Aversive conditioning in a model-mimic system. *Animal Behaviour* 35:75–80.

———, ed. 1988. *The ecology of social behavior.* San Diego, CA: Academic Press.

———. 2002. Cognition and communication in prairie dogs. In *The cognitive animal,* ed. Bekoff, M., C. Allen, and G. Burghardt, 257–264. Cambridge, MA: MIT Press.

Slobodchikoff, C. N., S. H. Ackers, and M. Van Ert. 1998. Geographical variation in alarm calls of Gunnison's prairie dogs. *Journal of Mammalogy* 79:1265–72.

Slobodchikoff, C. N., and R. Coast. 1980. Dialects in the alarm calls of prairie dogs. *Behavioral Ecology and Sociobiology* 7:49–53.

Slobodchikoff, C. N., J. Kiriazis, C. Fischer, and E. Creef. 1991. Semantic information distinguishing individual predators in the alarm calls of Gunnison's prairie dogs. *Animal Behaviour* 42:713–19.

Slobodchikoff, C. N., and J. Placer. 2006. Acoustic structures in the alarm calls of Gunnison's prairie dogs. *Journal of the Acoustical Society of America* 119:3153–60.

Slobodchikoff, C. N., A. Robinson, and C. Schaak. 1988. Habitat use by Gunnison's prairie dogs. In *Management of amphibians, reptiles, and small mammals in North America. General Technical Report RM-166,* ed. Szaro, R. C., K. E. Severson, and D. R. Patton, 403–8. Fort Collins, CO: USDA Forest Service, Rocky Mountain Forest and Range Experiment Station.

Smith, G. A., and M. V. Lomolino. 2004. Black-tailed prairie dogs and the structure of avian communities on the shortgrass plains. *Oecologia* 138:592–602.

Smith, G. W., and D. R. Johnson. 1985. Demography of a Townsend ground squirrel population in soutwestern Idaho. *Ecology* 66:171–78.

Smith, J. C. 2003. "Phil's Rescuers Hope to Make Clean Getaway." *Newsday,* August 23, A4.

Smith, J. L. S. 1982. Hibernation in the Zuni prairie dog. MS thesis. Flagstaff: Northern Arizona University.

Smith, R. E. 1967. Natural history of the prairie dog in Kansas. *Miscellaneous Publications of the Museum of Natural History, University of Kansas* 49:1–39.

Smith, W. J., S. L. Smith, E. C. Oppenheimer, and J. G. Devilla. 1977. Vocalizations of the black-tailed prairie dog, *Cynomys ludovicianus*. *Animal Behaviour* 25:152–64.

Snell, G. P., and B. D. Hlavacheck. 1980. Control of prairie dogs—the easy way. *Rangelands* 2:239–40.

Stapp, P. 1998. A reevaluation of the role of prairie dogs in the Great Plains grasslands. *Conservation Biology* 12:1253–59.

Stavins, R. 1991. *Project 88-Round II Incentives for action: Implementing market-based environmental policies and programs. A Public Policy Study sponsored by Senator Timothy E. Wirth, Colorado, and Senator John Heinz, Pennsylvania*, ed. R. Stavins. Washington, DC. On-line copy: http://ksghome.harvard.edu/~rstavins/Monographs_&_Reports/Project_88-2.pdf

Steinauer, E. M., and S. L. Collins. 1996. Prairie Ecology—The Tallgrass Prairie. In *Prairie Conservation: Preserving North America's most endangered ecosystem*, ed. F. Sampson, and F. Knopf, 339. Washington, DC: Island Press.

Steiner, A. L. 1974. Body-rubbing, marking, and other scent-relatd behavior in some ground squirrels (Sciuridae): A descriptive study. *Canadian Journal of Zoology* 52:889–906.

———. 1975. "Greeting" behavior in some sciuridae, from an ontogenetic, evolutionary, and socio-behavioral perspective. *Naturalist Canadien* 102:737–51.

Sterling, K. B. 2000a. Allen, Joseph Asaph. American National Biography Online, www.anb.org.

———. 2000b. Merriam, Clinton Hart. American National Biography Online, www.anb.org.

Stockard, A. H. 1929. Observations on reproduction in the white-tailed prairie dog *Cynomys leucurus*. *Journal of Mammalogy* 10:209–12.

Stromberg, M., R. Rayburn, and T. Clark. 1983. Black-footed ferret prey requirements: an energy balance estimate. *Journal of Wildlife Management* 47:67–73.

Stymne, S., and T. Jackson. 2000. Intra-generational equity and sustainable welfare: a time series analysis for the UK and Sweden. *Ecological Economics* 33:219–36.

Sullins, M. J., D. T. Theobold, J. R. Jones, and L. M. Burgess. 2002. Lay of the land: Ranch land and ranching. In *Ranching west of the 100th meridian*, ed. R. L. Knight, W. C. Gilgert, and E. Marston, 25–31. Washington, DC: Island Press.

Summers, C. A., and R. L. Linder. 1978. Food habits of the black-tailed prairie dog in western South Dakota. *Journal of Range Management* 31:134–36.

SWCS. 2000. Growing carbon: a new crop that helps agricultural producers and the climate too. Soil and Water Conservation Service, http://www.swcs.org/en/publications/books

Thomas, R. E., A. M. Barnes, T. J. Quan, M. L. Beard, L. G. Carter, C. E. Hopla. 1988. Susceptibility to *Yersinia pestis* in the Northern grasshopper mouse *(Onychomys leucogaster). Journal of Wildlife Diseases* 24:327–33.

Tileston, J. V., and R. R. Lechleitner. 1966. Some comparisons of the black-tailed and white-tailed prairie dogs in north-central Colorado. *American Midland Naturalist* 75:292–316.

Travis, S. E., and C. N. Slobodchikoff. 1993. Effects of food resource distribution on the social system of Gunnison's prairie dog *(Cynomys gunnisoni). Canadian Journal of Zoology* 71:1186–92.

Travis, S. E., C. N. Slobodchikoff, and P. Keim. 1995. Ecological and demographic effects on intraspecific variation in the social system of prairie dogs. *Ecology* 76:1794–1803.

———. 1996. Social assemblages and mating relationships in prairie dogs: A DNA fingerprint analysis. *Behavioral Ecology* 7:95–100.

———. 1997. DNA fingerprinting reveals low genetic diversity in Gunnison's prairie dog *(Cynomys gunnisoni). Journal of Mammalogy* 78:725–32.

Trevino-Villarreal, J., I. M. Berk, A. Aguirre, and W. E. Grant. 1998. Survey for sylvatic plague in the Mexican prairie dog *(Cynomys mexicanus). Southwestern Naturalist* 43:147–54.

Trudeau, K. M., H. B. Britten, and M. Restani. 2004. Sylvatic plague reduces genetic variability in black-tailed prairie dogs. *Journal of Wildlife Research* 40:205–11.

Truett, J. C. 2002. Aplomado falcons and grazing: Invoking history to plan restoration. *Southwestern Naturalist* 47:379–400.

Truett, J. C., J. L. D. Dullum, M. R. Matchett, E. Owens, and D. Seery. 2001. Translocating prairie dogs: A review. *Wildlife Society Bulletin* 29:863–72.

Tyler, J. D. 1968. Distribution and vertebrate associations of the black-tailed prairie dog in Oklahoma. PhD diss. Norman: University of Oklahoma.

Uresk, D. W. 1984. Black-tailed prairie dog food habits and forage relationships in western South Dakota. *Journal of Range Management* 37:325–29.

———. 1985. Effects of controlling black-tailed prairie dogs on plant production. *Journal of Range Management* 38:466–68.

Uresk, D. W., J. G. MacCracken, and A. J. Bjugstad. 1982. Prairie dog density and cattle grazing relationships. In *Fifth Great Plains Wildlife Damage Control Workshop Proceedings, 13–15 October 1982,* 199–201. Lincoln: University of Nebraska.

Uresk, D. W., and D. D. Paulson. 1988. Estimated carrying capacity for cattle competing with prairie dogs and forage utilization in western South Dakota. In *Management of amphibians, reptiles, and small mammals in North America. USDA. General Technical Report RM-166,* ed. R. C. Szaro, K. E. Severson, and

D. R. Patton, 387–90. Fort Collins, CO: USDA Forest Service, Rocky Mountain Forest and Range Experiment Station.

Uresk, D. W., and J. C. Sharps. 1986. Denning habitat and diet of the swift fox in western South Dakota. *Great Basin Naturalist* 46:249–53.

U.S. APHIS ADC (U.S. Animal, Plant, and Health Inspection Services; Division of Animal Damage Control). 1990. *Annual Report* Hyattsville, Maryland.

———. 1991. *Annual Report.* Hyattsville, Maryland.

USBIA. 1991. *Prairie dog control program, Cheyenne River and Rosebud Indian Reservations environmental assessment.* Aberdeen, South Dakota: United States Bureau of Indian Affairs.

U.S. Department of Agriculture. 1921. Death to rodents, by W. R. Bell. Yearbook No. 855:421–38. Washington, DC.

U.S. Department of Health and Human Services (HHS). Centers for Disease Control and Prevention (CDC) and Food and Drug Administration (FDA). 2003a. "Control of Communicable Diseases." *Federal Register* 68, no. 117: 36566–67. http://www.fda.gov/OHRMS/DOCKETS/98fr/061803c.htm.

———. Centers for Disease Control and Prevention (CDC) and Food and Drug Administration (FDA). 2003b. "Control of Communicable Diseases; Restrictions on African Rodents, Prairie Dogs, and Certain Other Animals." *Federal Register* 68, no. 213: 62353–69. http://www.fda.gov/OHRMS/DOCKETS/98fr/03-27557. htm.

———. Food and Drug Adminstration (FDA). 2003. "Economic Analysis and Monkeypox (2003N-0400)." FDA docket, December 28. http://google2.fda. gov/search?q=economic+analysis+and+monkeypox&client=FDA&site=docke ts&output=xml_no_dtd&lr=&proxystylesheet=FDA.

———. Food and Drug Administration (FDA). 2005. "African rodents and other animals that may carry the monkeypox virus." *Code of Federal Regulations* title 21, part 1240.63 (2005 edition). http://www.gpoaccess.gov/cfr/index.html.

U.S. Department of the Interior (DOI). 1995. *Federal and State Endangered Species Expenditures.* Washington, DC: U.S. Department of the Interior.

U.S. Fish and Wildlife Service. 1991a. *Endangered and threatened wildlife and plants.* Washington, DC: U.S. Department of the Interior.

———. 1991b. *Utah prairie dog recovery plan.* Denver, Colorado: USFW.

———. 2004. *Federal and State Endangered and Threatened Species Expenditures Fiscal Year 2004.* Washington, DC: U.S. Department of the Interior.

USNPS (United States National Park Service). 1992. Pesticide use logs, case incident reports, and miscellaneous documents 1982–1992. FOIA request to D. Roemer, 1993.

Van Pelt, W. E., ed. 1999. *The black-tailed prairie dog conservation assessment and strategy. Technical Report 159.* Phoenix: Nongame and Endangered Wildlife Program, Arizona Game and Fish Department.

————, ed. 2000. *The Arizona black-tailed prairie dog management plan. Draft 1.* Phoenix: Nongame and Endangered Wildlife Program, Arizona Game and Fish Department.

Van Putten, M., and S. D. Miller 1999. Prairie dogs: the case for listing. *Wildlife Society Bulletin* 27:1110–20.

Van Vuren, D., and M. P. Bray. 1983. Diets of bison and cattle on a seeded range in southern Utah. *Journal of Range Management* 36:499–500.

VerCauteren, T. L., S. W. Gillihan, and S. W. Hutchings. 2001. Distribution of burrowing owls on public and private lands in Colorado. *Journal of Raptor Research* 35:357–61.

Verdolin, J. L. 2007. Resources, not male mating strategies, determine social structure in Gunnison's prairie dogs. *Behaviour* 144: 1361–1382.

Verdolin, J. L., K. Lewis, and C. N. Slobodchikoff. 2008. Morphology of prairie dog burrow systems: a comparison among Gunnison's *(Cynomys gunnisoni)*, black-tailed *(Cynomys ludovicianus)*, white-tailed *(Cynomys leucurus)* and Utah *(Cynomys parvidens)* prairie dogs. *Southwestern Naturalist,* 53:201–7.

Verdolin, J. L., and C. N. Slobodchikoff. 2002. Vigilance and predation risk in Gunnison's prairie dogs. *Canadian Journal of Zoology* 80:1197–1203.

Vogel, S., C. P. Ellington, and D. L. Kilgore. 1973. Wind-induced ventilation of the burrow of the prairie dog, *Cynomys ludovicianus. Journal of Comparative Physiology* 85:1–15.

Vogt, K. A., J. C. Gordon, J. P. Wargo, D. J. Vogt, H. Asborjensen, P. A. Palmiotto, H. J. Clark, J. L. O'Hara, W. S. Keeton, T. Patel-Weynand, and E. Witten. 1997. Ecosystems: Balancing science with management. New York: Springer-Verlag.

Vogt, K. A., J. Oswald, K. H. Beard, J. L. O'Hara, and M. G. Booth. 2001. Conservation efforts, contemporary. *Encyclopedia of Biodiversity*, Volume 1. San Diego, CA: Academic Press.

Vosburgh, T. C., and L. R. Irby. 1998. Effects of recreational shooting on prairie dog colonies. *Journal of Wildlife Management* 62:363–71.

Wagner, D. M., and L. C. Drickamer. 2004. Abiotic habitat correlates of Gunnison's prairie dog in Arizona. *Journal of Wildlife Management* 68:188–97.

Wagner, D. M., L. C. Drickamer, D. M. Krpata, C. J. Allender, W. E. Van Pelt, and P. Keim. 2006. Persistence of Gunnison's prairie dog colonies in Arizona, USA. *Biological Conservation* 130:331–39.

Wallis, C. A. 1982. An Overview of the mixed grasslands of North America. In *Grassland Ecology and Classification Symposium Proceedings*, ed. A. C. Nicholson, A. McClean and T. E. Baker. Province of British Columbia: Ministry of Forests.

Waring, G. H. 1966. Sounds and communications of the yellow-bellied marmot *(Marmota flaviventris). Animal Behaviour* 14:177–83.

————. 1970. Sound communications of black-tailed, white-tailed, and Gunnison's prairie dogs. *American Midland Naturalist* 83:167–85.

Weaver, J. E., and F. W. Albertson. 1956. *Grasslands of the Great Plains*. Lincoln: Nebraska: Johnsen Publishing Co.

Weaver, T., E. M. Payson, and D. L. Gustafson. 1996. Prairie Ecology—The Shortgrass Prairie. In *Prairie Conservation: Preserving North America's most endangered ecosystem*, ed. F. Sampson and F. Knopf. Washington, DC: Island Press.

Weeks, W. W. 1997. *Beyond the Ark: tools for an ecosystem approach to conservation*. Washington, DC: Island Press.

Weltzin, J. F., S. Archer, and R. K. Heitschmidt. 1997. Small mammal regulation of vegetation structure in a temperate savannah. *Ecology* 78:751–63.

Whicker, A., and J. K. Detling. 1988. Ecological consequences of prairie dog disturbances. *Bioscience* 38:778–85.

Whicker, A. D., and J. K. Detling. 1989. Control of grassland ecosystem processes by prairie dogs. In *Symposium on the management of prairie dog complexes for the re-introduction of the black-footed ferret*, ed. J. L. Oldenmeyer, D. E. Biggins, B. J. Miller, and R. Crete, 18–27. Washington, DC: U. S. Department of Interior Fish and Wildlife Service.

White, G. C., J. R. Dennis, and F. M. Pusateri. 2005. Area of black-tailed prairie dog colonies in eastern Colorado. *Wildlife Society Bulletin* 33:265–72.

White, M. E., D. Gordon, J. D. Poland, and A. M. Barnes. 1980. Recommendations for the control of *Yersinia pestis* infections. *Inf. Contr.* 1:324–29.

Williams, J. R., and P. L. Diebel. 1996. The economic value of the prairie. In *Prairie conservation: preserving North America's most endangered ecosystem*, ed. F. Samson and F. Knopf. Washington, DC: Island Press.

Williams, R. J. 1982. The role of climate in a grassland classification. In *Grassland Ecology and Classification Symposium Proceedings*, ed. A. C. Nicholson, A. McClean, and T. E. Baker. Province of British Columbia: Ministry of Forests.

Wilson, E. O. 1992. *The diversity of life*. New York: Norton.

———. 1993. *The biophilia hypothesis*. Washington, DC: Island Press.

Wilson, E. O., and W. H. Bossert. 1971. *A primer of population biology*. Stamford, CT: Sinauer.

Winter, S. L., J. F. Cully, Jr, and J. S. Pontius. 2003. Breeding season avifauna of prairie dog colonies and non-colonized areas in shortgrass prairie. *Transactions of the Kansas Academy of Science* 106:129–38.

Woodward, R. T. 2000. Sustainability as intergenerational fairness: Efficiency, uncertainty, and numerical methods. *American Journal of Agricultural Economics* 82:581–93.

WRI. 1998. World Resources Institute, http://www.wri.org/biodiv/tropical.html.

Wright, H. A. 1974. Effect of fire on southern mixed prairie grasses. *Journal of Range Management* 27:417–19.

Wydeven, A. P., and R. B. Dahlgren. 1985. Ungulate habitat relationships in Wind Cave National Park. *Journal of Wildlife Management* 49:805–13.

Zinn, H. C., and W. F. Andelt. 1999. Attitudes of Fort Collins, Colorado, residents toward prairie dogs. *Wildlife Society Bulletin* 27:1098–1106.

Index

Index